W9-AUL-123

BUSINESS STATISTICS

DEMYSTIFIED

Bedford Free Public Library
Bedford, Massachusetts 01730

Home of the Bedford Flag

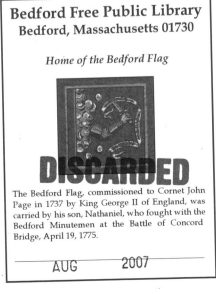

DISCARDED

The Bedford Flag, commissioned to Cornet John Page in 1737 by King George II of England, was carried by his son, Nathaniel, who fought with the Bedford Minutemen at the Battle of Concord Bridge, April 19, 1775.

AUG   2007

Demystified Series

BUSINESS STATISTICS

DEMYSTIFIED

STEVEN M. KEMP, Ph.D
SID KEMP, PMP

McGRAW-HILL
New York   Chicago   San Francisco   Lisbon   London
Madrid   Mexico City   Milan   New Delhi   San Juan
Seoul   Singapore   Sydney   Toronto

The McGraw·Hill Companies

Cataloging-in-Publication Data is on file with the Library of Congress

To Our Mom

Figure 7-16. Copyright © 1995–2004, The Weather Channel Interactive, Inc.

Copyright © 2004 by The McGraw-Hill Companies, Inc. All rights reserved. Printed in the United States of America. Except as permitted under the United States Copyright Act of 1976, no part of this publication may be reproduced or distributed in any form or by any means, or stored in a data base or retrieval system, without the prior written permission of the publisher.

3 4 5 6 7 8 9 0 DOC/DOC 0 9 8 7 6 5

ISBN 0-07-144024-0

The sponsoring editor for this book was Judy Bass and the production supervisor was Pamela Pelton. The art director for the cover was Margaret Webster-Shapiro. It was set in Times Roman by Keyword Publishing Services Ltd.

Printed and bound by RR Donnelley.

 This book was printed on recycled, acid-free paper containing a minimum of 50% recycled, de-inked fiber.

McGraw-Hill books are available at special quantity discounts to use as premiums and sales promotions, or for use in corporate training programs. For more information, please write to the Director of Special Sales, McGraw-Hill Professional, Two Penn Plaza, New York, NY 10121-2298. Or contact your local bookstore.

Information in this book has been obtained by The McGraw-Hill Companies, Inc. ("McGraw-Hill") from sources believed to be reliable. However, neither the publisher nor the authors guarantee the accuracy or completeness of any information published herein and neither McGraw-Hill nor its authors shall be responsible for any errors, omissions, or damages arising out of use of this information. This work is published with the understanding that McGraw-Hill and its authors are supplying information but are not attempting to render engineering or other professional services. If such services are required, the assistance of an appropriate professional should be sought.

CONTENTS

CONTENTS

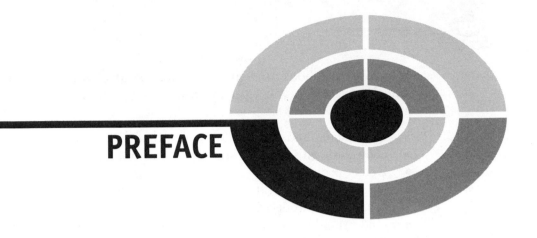

PREFACE

Many people find statistics challenging, but most statistics professors do not. As a result, it is sometimes hard for our professors and the authors of statistics textbooks to make statistics clear and practical for business students, managers, and executives. *Business Statistics Demystified* fills that gap. We begin slowly, introducing statistical concepts without mathematics. We build step by step, from defining statistics in Part One providing the basic tools for creating and understanding statistical reports in Part Two, introducing the statistical measures commonly—and some not-so-commonly—used in business in Part Three and, in Part Four, applying statistics to practical business situations with forecasting, quality management, and more.

Our approach is to focus on understanding statistics and how to use it to support business decisions. The math comes in when it is needed. In fact, most of the math in statistics is done by computers now, anyway. When the ideas are clear, the math will follow fairly easily.

Business Statistics Demystified is for you if:

- *You are in a business statistics class, and you find it challenging.* Whether you just can't seem to think like a statistician, or it's the math, or you're not sure what the problem is, the answer is here. We take you through all the rough spots step by step.
- *You are in a business statistics class, and you want to excel.* You will learn how to use statistics in real business situations, and how to prepare top-quality statistical reports for your assignments.
- *You are studying business statistics to move up the career ladder.* We show you where statistics can—and can't—be applied in practical business situations.

We wrote this book so that you would be able to apply statistics in a practical way. When you have finished with this book, you will find that you can:

- Understand and evaluate statistical reports
- Help perform statistical studies and author statistical reports
- Detect problems and limitations in statistical studies
- Select the correct statistical measures and techniques for making most basic statistical decisions
- Understand how to select the appropriate statistical techniques for making common business decisions
- Be familiar with statistical tools used in the most common areas of business
- Avoid all the most common errors in working with and presenting statistics
- Present effective statistical reports that support business decisions

HOW TO GET THE MOST OUT OF THIS BOOK

If you are just learning statistics, we recommend you start at the beginning, and work your way through. We demystify the things that other books jump over too quickly, leaving your head spinning. In fact, you might read Part One before you look at other books, so you càn avoid getting mystified in the first place!

If you are comfortable with statistics, skim Part One and see if it clarifies some of the vague ideas we can all carry around without knowing it, and then use the rest of the book as you see fit. If you want to focus on performing statistical studies and preparing statistical reports—or even just reading them—then Part Two will be a big help. Part Three is a useful reference for the more advanced statistical techniques used in business. And Part Four makes the link between statistics and business interesting and exciting.

SIDEBARS FOR EASY LEARNING

In *Business Statistics Demystified*, we want to make it easy for you to learn and to find what you need to know. So we've created several different types of sidebars that will introduce key ideas. Here they are:

- *Tips on Terms*. Definitions and crucial terminology.
- *Critical Cautions*. Something statistical you must do—or must avoid—to get things right.

- *Study Review.* Key points for exam preparation.
- *Survival Strategies.* What to do on the job.
- *Handy Hints.* Other practical advice.
- *Fun Facts.* A little bit on the lighter side.
- *Case Studies.* Real-world examples that teach what works—and what doesn't.
- *Bio Bites.* The authors' experience—if you learn from what we've been through, your statistical work will be easier.
- *Quick Quotes.* Bits of wisdom from folks much smarter than we are.

ACKNOWLEDGMENTS

Our first thanks go to Scott Hoffheiser, our administrative assistant, whose understanding of statistics, proofreading skill, and skills with Microsoft Equation Editor® and in creating graphs with Microsoft Excel® were indispensable, and are well illustrated in *Business Statistics Demystified*. If you like the quizzes, then you will be as grateful as we are to Anna Romero, Ph.D. Professor Mark Appelbaum, currently of the University of California, San Diego, was the first person to be successful in teaching me (Steve) statistics and deserves special thanks for that. Our Dad, Bernie Kemp, a now retired professor of economics, offered some wonderful suggestions, which improved the book immensely. More importantly, he taught us about numbers before we learned them in school. Most importantly, we learned all about the uncertainty of the world and the limits of measurement at his knee. Our Mom, Edie Kemp, provided support, which allowed us the time to write, always the sine qua non of any book, as did Kris Lindbeck, Sid's wife.

Dave Eckerman and Peter Ornstein, both of the Psychology Department at the University of North Carolina at Chapel Hill, have supported the first author's affiliation with that institution, whose extensive research resources were invaluable in the preparation of the manuscript of the book.

PART ONE

What Is Business Statistics?

People in business want to make good decisions and implement them. When we do, our businesses flourish, we solve problems, we make money, we succeed in developing new opportunities, etc. In the work of implementation—executing business plans—statistics can't play much of a part. But in the making of good decisions—in planning, choosing among options, finding out what our customers, our manufacturing plants, or our staff are thinking and doing, and controlling the work of people and machinery—business people need all the help we can get. And statistics can help a great deal.

To understand how statistics can help business people understand the world, it is important to see the bigger picture, of which business statistics is a part. This is illustrated in Fig. I-1.

Let's start at the top. Philosophy is the field that asks, and tries to answer, questions that folks in other fields take for granted. These include questions like: What is business? What is mathematics? How can we relate mathematics to science, engineering, and statistics? We left out the arrows because

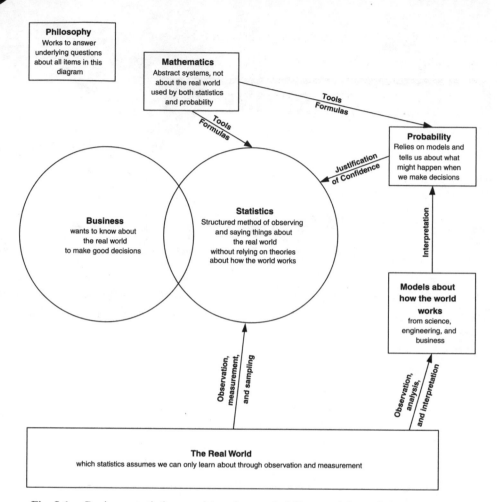

Fig. I-1.   Business statistics, mathematics, probability, models, and the real world.

philosophy takes every other field as its field of study. And the first piece of good news is that, while the authors of a good statistics book may need to worry about philosophy, you don't.

Next, mathematics can't help business directly, because it is a pure abstraction, and business people want to understand, make decisions about, work in, and change the real world. Statistics brings the power of mathematics to the real world by gathering real-world data and applying mathematics to them. The second piece of good news is that, while statistics often uses mathematics, statisticians often don't need to. In the practical world of business statistics, we leave the math (or at least the calculations) to

computers. But we do need to understand enough math to:

- understand the equations in statistical tools,
- know which equations to use when, and
- pass the exams in our statistics classes.

QUICK QUOTE

All real statistics can be done on an empty beach drawing in the sand with a stick. The rest is just calculation.

John Tukey

The next point is key: statistics is not a part of mathematics. It is its own field, its own discipline, independent of math or other fields. But it does make use of mathematics. And it has important links to science, engineering, business models of the world, and probability.

KEY POINT

Statistics Stands by Itself
Statistics is not part of mathematics, probability, business, science, or engineering. It stands independent of the others. At the same time, statistics does make use of, and relate to, mathematics, probability, science, and engineering. And it can help business people make good decisions.

A fundamental problem of business—perhaps the fundamental problem of life—is that we would love to know exactly how the world works and know everything that is going on, but we can't. Instead, we have only partial information—all too often inaccurate information—about what is going on in the real world. We also have a bunch of guesses—often called theories, but we will call them models—about how the real world works. The guesses we use in business often come from experts in science, engineering, the social sciences, and business theory.

When business executives turn to experts for help in making decisions, we often run into a problem. We understand that the experts know their stuff. But what if their whole model is wrong? The most we can give to anyone coming to us with a model of how the world works—a business model, a scientific model, a social model, or an engineering model—is this: If your model is right, then your advice will improve my chances of getting me to the right decision. But what if your model is wrong?

In this, statistics stands apart from other fields. Engineering, science, the social sciences, and business models all rely on being right about how the world works. Statistics does not. Statistics relies on only one basic assumption: that the future is likely to resemble the past, in general. If we accept that principle, we can use statistics to understand the world, even if we have no model about how the world works, or no model we are confident enough to use.

Part of making good decisions is avoiding assumptions that might be wrong. In using statistics, we are avoiding the assumption that a particular idea of how the world works—how customers look at advertisements, or how vendors deliver their goods—is true. We are relying on more general, more proven principles.

But we can't use statistics for every business decision. And *Business Statistics Demystified* will show you how to know when statistics can help with business decisions, how to use good statistics, and how to spot and avoid bad statistical methods and unreliable statements backed up with a lot of good-sounding statistics. Also, parts of statistical theory, especially those regarding the significance of statistical results, were invented for statistics in its relationship to science. Determining what statistical results mean for business is very different from deciding what statistical results are important for science, and we will demystify that, as well.

Statistics helps business in two basic ways. The first is called *descriptive statistics,* and it tells us some useful things about what is going on in the data we have about the world. The second is called *inferential statistics,* and it helps us know about things we can't affordably measure and count, and about what is likely to happen if we make a particular decision.

We open *Business Statistics Demystified* with three chapters that lay a foundation for the rest of the book. Chapter 1 "Statistics for Business" expands on and clarifies the issues we have raised here: What is statistics, and how does it help us make business decisions? We also explore the basis of statistics, and explain why knowing how to do bad statistics is essential to not being fooled by them, and also for doing good statistics.

In Chapter 2 "What Is Statistics?" you will learn the basic elements and terms of statistics: Measurement, error, sampling, and analyzing. In Chapter 3 "What Is Probability?" we briefly turn away from statistics to introduce a related, but separate field, probability. Probability and statistics seem similar. Both apply mathematics to the real world. Both try to tell us what we are likely to find in the real world, or what is likely to happen if we make a certain decision. But there is a fundamental difference. Probability is a way of relating models to the real world and statistics is a way of finding out about the world without models. We will then distinguish probability from statistics. Finally, we will also show how the two work together to help us have confidence in our methods and decisions.

When we make the right decisions, and have confidence in them, it is easier to follow through on them. And when we make the right decision and follow through, we solve problems and succeed.

CHAPTER 1

Statistics for Business

Statistics is the use of numbers to provide general descriptions of the world. And business is, well, business. In business, knowing about the world can be very useful, particularly when it comes to making decisions. Statistics is an excellent way to get information about the world. Here, we define business statistics as the use of statistics to help make business decisions.

In this chapter, we will learn what statistics is for and how it ties into business. We will discuss generally what statistics can and cannot do. There will be no math and almost no technical terminology in this chapter (there will be plenty of time for that later). For now, we need to understand the basics.

Doing Without Statistics

Statistics is like anything else in business. It should be used only if it is worthwhile. Using statistics takes time, effort, and resources. Statistics for its own sake just lowers profits by increasing expenses. It is extremely important to recognize when and where statistics will aid in a business decision.

Business decisions, big and small, get made every day without statistics. The very smallest decisions will almost never benefit from statistics. What restaurant to take our client to for lunch is probably a decision best made without statistical assistance. There are many reasons not to use statistics for bigger decisions as well. Statistics is one of the most effective ways to convert specific facts about the world into useful information, but statistics cannot improve the quality of the original facts. If we can't get the right facts, statistics will just make the wrong facts look snazzy and mathematical and trustworthy. In that case, statistics may make us even worse-off than if we hadn't used them at all. It is vital to understand what facts are needed in order to make a good decision *before* we use statistics, and even before we decide what statistics to use.

KEY POINT

Facts First!
For example, if you are planning to take a foreign business guest to an excellent restaurant, you might think it's a good idea to pick the best restaurant in Chicago. Looking at customer surveys, that's likely to be a steak house.

But the more relevant information might be the fact that your guest is a vegetarian.

The lesson: Decide what's important, get the right facts, and then do statistics if they help.

Even if the facts are right, there may not be enough of them to help us make our decision. If so, the general information we get from statistics will not be precise or accurate enough for our needs. In statistics, imprecision and inaccuracy are called *error*. Error is one of the most important aspects of statistics. One of the most remarkable things about statistics is that we can use statistics to tell us how much error our statistics have. This means that sometimes we can use statistics to find out when not to use statistics.

Statistics are Cheap

Is statistics overused or underused in business? It is hard to say. Some business decisions are not made using statistics and some business decisions should not be. But deciding when to use statistics is often not easy. Many business decisions that could use statistical information are made without statistics and many business decisions that shouldn't use statistics are made using statistics. It is probably fair to say that there are types of decisions and areas of business where statistics are underused and others where they are overused.

Things that lead to the underuse of statistics are:

- lack of statistical knowledge on the part of the business people
- mistaken assumptions about how complicated or difficult to use or costly statistics can be
- the time pressure to make business decisions
- a failure to set up statistical systems in advance of decision making

Your decision to learn about statistics will help you avoid the underuse of statistics and *Business Statistics Demystified* will help you do that.

Things that lead to the overuse of statistics are:

- requirements made by bosses, standards, organizations, and legal authorities that fail to recognize the limitations of statistics
- failures by decision makers to determine the value of statistics as part of their analysis
- a poor understanding of the limits of the available facts or the statistical techniques useful for converting those facts into information
- a desire to justify a decision with the appearance of a statistical analysis

Learning about statistics means more than learning what statistics is and what it can do. It means learning about how numbers link up to the world and about the limits of what information can be extracted. This is what it means to think statistically. Far more important than learning about the specific techniques of statistics is learning how to think statistically about real business problems. This book will help you do both.

Lying with Statistics

There is a wonderful book by Huff and Geis (1954) called *How to Lie with Statistics*. In clear and simple terms, it shows how statistics can be used to misinform, rather than inform. It also provides wonderful examples about how to think statistically about problems and about how to read statistical information critically. (If *How to Lie with Statistics* covered all of basic statistics and was focused on business, there might be no need for this book!) The real importance of knowing how to lie with statistics is that it is the best way to learn that careful, sound judgment is vital in making statistics work for us while making business decisions. Identifying a problem and applying the formulas without understanding the subtleties of how to apply statistics to business situations is as likely to hurt our decision making as it is to help it.

KEY POINT

97% Fat-Free

The term fat-free on food labels is an excellent example of what we mean by lying with statistics. It would be easy to think that 97% fat-free meant that 97% of the original fat had been removed. Not at all. It means that 3% of the milk is fat. So 97% fat-free just means "3% fat." But how well would that sell?

There are two lessons here: First, we can only build good statistics if we gather and understand all the relevant numbers. Second, when we read statistical reports—on our job or in the newspaper—we should be cautious about incomplete measurements and undefined terms.

Each and every statistical measure and statistical technique have their own strengths and limitations. The key to making statistics work for us is to learn those strengths and limitations and to choose the right statistics for the situation (or to choose not to use statistics at all when statistics cannot help). Throughout this book, we will learn about each statistical measure and technique in terms of what it can and cannot do in different business situations, with respect to different business problems, for making different business decisions. (We will also slip in the occasional fun example of how statistics get misused in business.)

So Many Choices, So Little Time

One feature of statistics is the enormous number of widely different techniques available. It is impossible to list them all, because as we write the list, statisticians are inventing new ones. In introducing statistics, we focus our attention on the most common and useful statistical methods. However, as consumers of statistics and statistical information, we need to know that there are lots more out there. Most often, when we need more complicated and sophisticated statistics, we will have to go to an expert to get them, but we will still have to use our best statistical judgment to make sure that they are being used correctly.

Even when we are choosing from basic statistical methods to help with our business decisions, we will need to understand how they work in order to make good use of them. Instead of just memorizing the fact that medians should be used in measuring salaries and means should be used in measuring monthly sales, we need to know what information the median gives us that the mean does not, and vice versa. That way, when a new problem shows up in our business, we will know what statistic to use, even if it wasn't on a list in our statistics book.

When we get past basic statistical measures and onto basic statistical techniques, we will learn about *statistical assumptions*. Each statistical technique has situations in which it is guaranteed to work (more or less). These situations are described in terms of assumptions about how the numbers look. When the situation we face is different than that described by the assumptions, we say that the assumptions *do not hold*. It may still work to use the statistical technique when some of the assumptions do not hold, but we have lost our guarantee. If there is another statistical technique that we can use, which has assumptions closer to the situation we are actually in, then we should consider using that technique instead.

CRITICAL CAUTION

Whenever a statistical technique is taught, the assumptions of that technique are presented. Because the assumptions are key to knowing when to apply one technique instead of another, it is vitally important to learn the assumptions along with the technique.

One very nice thing about statistical assumptions is that, because they are written in terms of how the numbers look, we can use statistics to decide whether the statistical assumptions hold. Not only will statistics help us with our business decisions, but we will find that statistics can often help us with the statistical decisions that we need to make on the way to making our business decisions.

In the end, it is just as important to know how to match the type of statistics we use to the business decision at hand as it is to know how to use each type of statistic. This is why every statistics book spends so much time on assumptions, as will we.

Math and Mystery

Now comes the scary part: math. As we all have heard over and over again, mathematics has become part of our everyday life. (When I was a kid, computers were big things in far-off places, so we didn't believe it much. Now that computers are everywhere, most people see how math has taken over our world.) Up to a certain point, the more you understand math, the better off you are. And this is true in business as well.

But math is only a part of our world when it does something useful. Most of the mathematics that a mathematician worries about won't bother us in our world, even in the world of business. Even understanding all the math won't be especially helpful if we don't know how to apply it. Statistics is a very odd subject, in a way, because it works with both abstract things like math, and with the very real things in the world that we want to know about. The key to understanding statistics is not in understanding the mathematics, but in understanding how the mathematics is tied to the world. The equations are things you can look up in a book (unless you are taking an exam!) or select off a menu in a spreadsheet. Once you understand how statistics links up numbers to the world, the equations will be easy to use.

Of course, this does not mean that you can get by without the algebra required for this book (and probably for your statistics class). You need to understand what a constant is, what a variable is, what an equation is, etc. If you are unsure of these things, we have provided Appendix A with some of the basic definitions from algebra.

Where Is Statistics Used?

At the start of this chapter, we defined *business statistics* as statistics used to help with business decisions. In business, decisions are everywhere, little ones, big ones, trivial ones, important ones, and critical ones. As the quotation by Abraham Lincoln suggests, the more we know about what is going on, the more likely we are to make the right decision. In the ideal, if we knew specifics about the future outcome of our decision, we would never make a mistake. Until our boss buys us a crystal ball so that we can see into the future, we will have to rely on using information about the present.

QUICK QUOTE

If we could first know where we are, and whither we are tending, we could better judge what to do, and how to do it.

Abraham Lincoln

But what sort of information about the present will help us make our decision? Even if we know everything about what is going on right now, how do we apply that information to making our decision? The simple answer is that we need to look at the outcomes of similar decisions made previously in similar circumstances. We cannot know the outcome of our present decision, but we can hope that the outcomes of similar decisions will be similar.

The central notion of all statistics is that similar past events can be used to predict future events. First and foremost, this assumption explains why we have defined statistics as the use of numbers to describe *general* features of the world. No specific fact will help us, except for the specific future outcome of our decision, and that is what we can't know. In general, the more we know about similar decisions in the past and their results, the better we can predict the outcome of the present decision. The better we can predict the outcome of the present decision, the better we can choose among the alternative courses of action.

FUN FACTS

The statistical notion that past events can be used to predict future ones is derived from a deeper philosophical notion that the future will be like the past. This is a central notion to all of Western science. It gives rise to the very famous "Humean dilemma" named after the philosopher, David Hume, who was the first person to point out that we cannot have any evidence that the future will be like the past, except to note that the future has been like the past in the past. And that kind of logic is what philosophers call a *vicious circle*.

We discuss this problem more deeply in Chapter 16 "Forecasting."

There are three things we need to know before statistics can be useful for a business decision. First, we need to be able to characterize the current decision we face precisely. If the decision is to go with an ad campaign that is either "edgy" or "dynamic," we will need to know a lot about what is and is not an edgy or a dynamic ad campaign before we can determine what information about past decisions will be useful. If not, our intuition, unassisted by statistics, may be our best bet. It is also important to be able to determine what general features of the world will help us make our decision. Usually, in statistics, we specify what we need to know about the world, by framing a question about general characteristics of the world as precisely as possible. And, of course, we don't need to describe the whole world. In fact, defining which part of the world we really need to know about is a key step in deciding how to use statistics to help with our decisions. For example, if we are predicting future sales, it is more valuable to know if our company's specific market is growing than to know if the general economy is improving. We'll look at these issues further in Part Four, when we discuss forecasting.

Second, there needs to be a history of similar *situations* that we can rely upon for guidance. Happily, here we are assisted by nature. Wildly different situations have important features in common that we can make use of in statistics. The important common elements can be found and described by abstracting away from the details of the situation, using numbers. This most important concept of abstraction is very simple and we have a lot of experience with it. We all learned very early on that, once we learned to count marbles and pencils we could also count sheep, cars, and dollars.

When we think about what we've done, we realize that we've defined a new practice, counting, and created a new tool for understanding the world, the count. The number of pennies in a jar or the number of sheep in a flock is not a specific fact about one specific penny or sheep. It is a general fact about the contents of the jar or the size of the flock. A *count* is a statistical measure that we use to tell us the quantity we have of an item. It is the first and simplest of what are called *descriptive statistics,* since it is a statistical measure used to describe things.

If our general question about the world merely requires a description of the current situation or of previous similar situations as an answer, descriptive statistics may be enough. Examples of questions that call for descriptive statistics are:

- How many married women between 18 and 34 have purchased our product in the past year?
- How many of our employees rate their work experience as very good or excellent?
- Which vendor gave us the best price on our key component last quarter?
- How many units failed quality checks today?
- How many consumers have enough disposable income to purchase our premier product?

Third, there needs to be a history of similar *decisions* that we can rely upon for guidance. While descriptive statistics have been around in some form since the beginning of civilization and the serious study of statistics has been around for almost a thousand years, it has been less than a hundred years since statisticians figured out how to describe entire decisions with numbers so that techniques useful in making one decision can be applied to other, similar decisions. The techniques used are at the heart of what is called *inferential statistics*, since they help us reason about, or make inferences from, the data in a way that provides answers, called conclusions, to our precisely phrased questions. In general, inferential statistics answers questions about *relations* between general facts about the world. The answers are based not only on relationships in the data, but also on how relationships of that same character can have an important effect on the consequences of our decisions.

If our question about the world requires a conclusion about a relationship as an answer, inferential statistics may be able to tell us, not only if the relationship is present in the data, but if that relationship is strong enough to give us confidence that our decision will work out. Examples of questions that call for inferential statistics are:

- Have men or women purchased more of our product in the past year?
- Do our employees rate their work experience more highly than do our competitors' employees?
- Did our lowest priced vendor give us enough of a price break on our key component last quarter to impact profits?
- Did enough units fail quality checks today to justify a maintenance call?
- How many consumers have enough disposable income to purchase our premier product if we lower the price by a specific amount?

TIPS ON TERMS

Descriptive statistics.   Statistical methods, measures, or techniques used to summarize groups of numbers.

Inferential statistics.   Statistical methods, measures, or techniques used to make decisions based groups of numbers by providing answers to specific types of questions about them.

Using statistics to make decisions in business is both easier and harder than using statistics in the rest of life. It is easier because so much of a business situation is already described with numbers. Inventories, accounts, sales, taxes, and a multitude of other business facts have been described using numbers since ancient Sumeria, over 4000 years ago. It is harder because, in business, it is not always easy to say what makes the best decision best. We may want to increase profits, or market share, or saturation, or stock price, etc. As we will see in Part Four, it is much easier to use statistics to predict the immediate outcome of our decision than it is to know if, in the end, it will be good for business.

CASE STUDY

Selling to Men and Women

For example, say that we know that more women than men bought our product during the Christmas season. And we know that, statistically, more women between 18 and 34 bought our product than the competitors'. Does that tell us whether we should focus our advertising on men or women in the spring? Not necessarily. It depends on whether we are selling a women's perfume or a power tool.

If perfume, maybe we should focus on men to buy Valentine's Day gifts. Or maybe on women, so they'll ask their husbands and boyfriends for our perfume by name.

If a power tool, then the Christmas season sales might be gifts. And a spring advertisement might be better focused on men who will be getting ready for summer do-it-yourself projects.

The lesson: Statistics may or may not be valuable to business. Common sense always is. If we use statistics, be sure to use them with some common sense thrown in.

CRITICAL CAUTION

Good statistics is not just a matter of knowing how to pick the techniques and apply them. Good statistics means knowing what makes for the best outcome and what the problems are in measuring the situation. Good business statistics demands a good understanding of business.

The Statistical Study

While statistics can be used on a one-time-only basis to help make a single business decision, most commonly we find that a *statistical study*, containing many statistics, either descriptive, or both descriptive and inferential, is conducted. The reason for this is that, when many decisions have to be made for one company, or for one department, or one project, and so forth, the situations that must be studied to make good choices for each decision may have a lot in common. A single statistical study can collect and describe a large amount of information that can be used to help make an even larger number of decisions. Like anything else, the economies of scale apply to statistics. It is much cheaper to collect a lot of statistics all at once that may help with lots of decisions later on than to collect statistics one by one as they are needed. In fact, as we will see later, both governmental agencies and private firms conduct statistical studies containing thousands of statistics they have no use for, but will be of use (and value) to their customers. We will have much to say about statistical studies in Part Two.

TIPS ON TERMS

Statistical study. A project using statistics to describe a particular set of circumstances, to answer a collection of related questions, or to make a collection of related decisions.

Statistical report. The document presenting the results of a statistical study.

The Statistical Report

No less important than the statistical study is the reporting of the results. Too often we think of statistics as the collection of the information and the calculation of the statistical measures. No amount of careful data collection or clever mathematics will make up for a statistical report that does not make the circumstances, assumptions, and results of the study clear to the audience. Statistics that cannot be understood cannot be used. One of the most important goals of this book is to explain how to read and understand a statistical report. Another equally important goal is to show how to create a report that communicates statistics effectively.

The rules for effective communication of statistics include all the rules for effective communication in general. Presenting numbers clearly is difficult to begin with, because much of our audience is not going to be comfortable with them. One solution is to present the numbers pictorially, and different kinds of numbers require different kinds of pictures, charts, and graphs. In addition, the numbers that result from statistical calculations are meaningful only as they relate to the business decisions they are intended to help. Whether we present them as numbers or as pictures, we need to be able to present them so that they are effective in serving their specific purpose.

Quiz

1. What do we call the use of numbers to provide general descriptions of the world to help make business decisions?
   (a) Common sense
   (b) Statistics
   (c) Business statistics
   (d) Mathematics

2. Which of the following does *not* lead to the underuse of statistics in business?
    (a)   A failure to set up statistical systems in advance of decision making
    (b)   A poor understanding of the limits of the available facts or the statistical techniques useful for converting those facts into information
    (c)   Lack of statistical knowledge on the part of business persons
    (d)   The time pressure to make business decisions

3. Which of the following does *not* lead to the overuse of statistics in business?
    (a)   Mistaken assumptions about how complicated or difficult to use or costly statistics can be
    (b)   Requirements made by bosses and standards organizations and legal authorities that fail to recognize limitations of statistics
    (c)   A desire to justify a decision with the appearance of a statistical analysis
    (d)   Failures by decision makers to determine the value of statistics as a part of their analysis

4. The key to knowing when to apply one statistical technique instead of another is to understand the _____ of the techniques.
    (a)   Error
    (b)   Statistical assumptions
    (c)   Mathematics
    (d)   History

5. Which of the following is *not* one of the three things that we need to know, and can know, before statistics can be useful for a business decision?
    (a)   We need to be able to characterize the current decision we face precisely
    (b)   There needs to be a history of similar situations that we can rely upon for guidance
    (c)   We need to know specific facts about the future outcome of our decision
    (d)   There needs to be a history of similar decisions that we can rely upon for guidance

6. Which of the following is a question that can adequately be answered by descriptive statistics?
    (a)   How many units failed quality checks today?
    (b)   Did our lowest priced vendor give us enough of a price break on our key component last quarter to impact profits?
    (c)   Have men or women purchased more of our product in the past year?
    (d)   Do our employees rate their work experience more highly than do our competitors' employees?

7. Which of the following is a question that can adequately be answered by inferential statistics?
    (a) How many of our employees rate their work experience as very good or excellent?
    (b) How many women between 18 and 34 have purchased our product in the past year?
    (c) Which vendor gave us the best price on our key component last quarter?
    (d) Did enough units fail quality checks today to justify a maintenance call?

8. What are the advantages of conducting a statistical study over using a statistical technique on a one-time only basis?
    (a) It is cheaper to collect a lot of statistics at once that may help with a lot of decisions later on than to collect statistics one by one as they are needed
    (b) A single statistical study can collect and describe a large amount of information that can be used to help make an even larger number of decisions
    (c) Both (a) and (b) are advantages
    (d) Neither (a) nor (b) are advantages

9. Which of the following components of a statistical study is not necessary to present in a statistical report?
    (a) The calculations of the statistical techniques used in the statistical study
    (b) The circumstances of the statistical study
    (c) The assumptions of the statistical study
    (d) The results of the statistical study

10. Which of the following is not an advantage of understanding how to lie with statistics?
    (a) It is the best way to learn that sound judgment is vital to making statistics work for us
    (b) It allows us to create convincing advertising campaigns
    (c) It helps us to learn the strengths and limitations of statistical measures and techniques
    (d) It helps us to be cautious about incomplete measurements and undefined terms in statistical reports

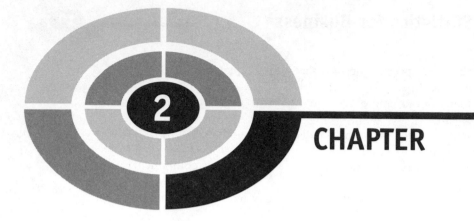

CHAPTER

2

What Is Statistics?

We have learned what it is that statistics does, now we need to find out a bit about how it works. How do statistical measures describe general facts about the world? How do they help us make inferences and decisions? There is a general logic to how statistics works and that is what we will learn about here. There will be no equations in this chapter, but we will introduce and define important technical terms.

SURVIVAL STRATEGIES

Use the definition sidebars and the quizzes to memorize the meaning of the technical terms in this chapter. The more familiar and comfortable you are with the terminology, the easier it will be to learn statistics.

This chapter will cover four very important topics: measurement, error, sampling, and analysis. Sampling, measurement, and analysis are the first three steps in doing statistics. First, we pick what we are going to measure, then we measure it, then we calculate the statistics.

We have organized the chapter so that the basic concepts are presented first and the more complicated concepts that require an understanding of the more basic concepts are presented afterwards. This will allow us to introduce most of the basic statistical terminology used in the rest of the book. But it will mean presenting these topics out of order compared to the order they are done in a statistical study.

These four topics relate to one another as follows: We need to measure the world to get numbers that tell us the details and then do statistical analysis to convert those details into general descriptions. In doing both measurement and analysis, we inevitably encounter error. The practice of statistics involves both the acknowledgment that error is unavoidable and the use of techniques to deal with error. Sampling is a key theoretical notion in understanding how measurements relate to the world and why error is inevitable.

Measurement

Statistics is not a form of mathematics. The most important difference is that statistics is explicitly tied to the world. That tie is the process of measurement.

WHAT IS MEASUREMENT?

The first and most fundamental concept in statistics is the concept of *measurement*. Measurement is the process by which we examine the world and end up with a description (usually a number) of some aspect of the world. The results of measurement are *specific* descriptions of the world. They are the first step in doing statistics, which results in *general* descriptions of the world.

Measurement is a formalized version of observation, which is how we all find out about the world every day. Measurement is different from ordinary day-to-day observation because the procedures we use to observe and record the results are specified so that the observation can be repeated the same way over and over again.

When we measure someone's height, we take a look at a person; apply a specific procedure involving (perhaps) a measuring tape, a pencil, and a part of the wall; and record the number that results. Let's suppose that we measure Judy's height and that Judy is "five foot two." We record the number 62, measured in inches. That number does not tell us a lot about Judy. It just tells us about one aspect of Judy, her height. In fact, it just tells us about her height on that one occasion. (A few years earlier, she might have been shorter.)

Statistics uses the algebraic devices of *variables* and *values* to deal with measurements mathematically. In statistics, a variable matches up to some aspect of the thing being measured. In the example above, the variable is height. The value is the particular number resulting from the measurement on this occasion. In this case, the value is 62. The person who is the *subject* of the measurement has many attributes we could measure and many others we cannot. Statisticians like to think of subjects (whether they are persons or companies or business transactions) as being composed of many variables, but we need to remember that there is always more to the thing being measured than the measurements taken. A person is more than her height, weight, intelligence, education level, occupation, hair color, salary, and so forth. Most importantly, not every variable is important to every purpose on every occasion. There are always more attributes than there are measurable variables, and there are always lots more variables that can be measured than we will measure.

KEY POINT

Vital to any statistical analysis will be determining which variables are relevant to the business decision at hand. The easiest things to measure are often not the most useful, and the most important things to know about are often the hardest to measure. The hardest part of all is to determine what variables will make a difference in making our business decision.

TIPS ON TERMS

Subject. The individual thing (object or event) being measured. Ordinarily, the subject has many attributes, some of which are measurable features. A subject may be a single person, object, or event, or some unified group or institution. So long as a single act of measuring can be applied to it, it can be considered a single subject. Also called the "unit of analysis" (not to be confused with the unit of measurement, below).

Occasion. The particular occurrence of the particular act of measurement, usually identified by the combination of the subject and the time the measurement is taken.

Situation. The circumstances surrounding the subject at the time the measurement is taken. Very often, when multiple measurements of a subject are taken on a single occasion, measurements characterizing the situation are also taken.

Value. The result of the particular act of measurement. Ordinarily, values are numbers, but they can also be names or other types of identifiers. Each value usually describes one aspect or feature of the subject on the occasion of the measurement.

Variable. A mathematical abstraction that can take on multiple values. In statistics, each variable usually corresponds to some measurable feature of the subject. Each measurement usually results in one value of that variable.

Unit. (Short for unit of measurement. Not to be confused with unit of analysis in the definition of Subject, above.) For some types of measurement, the particular standard measure used to define the meaning of the number, one. For instance, inches, grams, dollars, minutes, etc., are all units of measurement. When we say something weighs two and a half pounds, we mean that it weighs two and a half times as much as a standard pound measure.

Data. The collection of values resulting from a group of measurements. Usually, each value is labeled by variable and subject, with a timestamp to identify the occasion.

Values that aren't numbers

In statistics, measurement doesn't always result in numbers, at least not numbers in the usual sense. Suppose we are doing an inventory of cars in a car lot. We want to make a record of the important features of each car: make, model, year, and color. (Afterwards, we may want to do some statistics, but that can wait for a later chapter.) Statisticians would refer to the process of collecting and recording the make, model, year, and color of each car in the lot as measurement, even though it's not much like using a tape measure or a scale, and only in the case of the year does it result in a number. The reason for this is that, just like measuring height or weight, recording the color of an automobile results in a description of one feature of that particular car on that particular occasion. From a statistical point of view, the important thing is not whether the result is a number, but whether the results, each of which is a specific description of the world, can be combined to create general descriptions of the world. In the next section, *Levels of Measurement*, we will see how statisticians deal with non-numerical values.

TIPS ON TERMS

Categorical data. Data recorded in non-numerical terms. It is called categorical because each different value (such as car model or job title) places the subject in a different category.

Numerical data. Data recorded in numerical terms. There are different types of numerical data depending upon what numbers the values can be. (See *Levels of Measurement* below.)

What is data?

In Chapter 1 "Statistics for Business," we didn't bother too much about specific definitions. Now, in Chapter 2 "What is Statistics?"we are starting to concern ourselves with more exact terminology. Throughout the remainder of the book, we will try to be as consistent as possible with our wording, in order to keep things clear. This does not mean that statisticians and others who use statistics are always as precise in their wording as we should be. There is a great deal of confusion about certain terms. Among these are the notorious terms, *data* and *information*.

The values recorded as the result of measurement are data. In order to distinguish them from other sorts of values, we will use the term *data values*. Data are not the facts of the world that were measured. Data are descriptions, not the things described. Data are not the statistical measures calculated from the data values, no matter how simple. Often, statisticians will distinguish between "raw" data and "cleaned" data. The raw data are the values as originally recorded, before they are examined and edited. As we will see later on, cleaning data may involve changing it, but does not involve summarizing it or making inferences from it.

QUICK QUOTE

The map is not the territory.

Alfred Korzybski

KEY POINT

Data are specific descriptions. Statistics are general descriptions.

A lot of data is used only indirectly, in support of various statistical techniques. And data are always subject to error. To the degree that data contain error, they cannot inform. So data, even though they are information in the informal computer science sense, contain both information and error in the more technical, theoretical sense. In statistics, as in information theory, it is this latter, more technical sense that is most important. Because we will be using data to make business decisions, we must not forget that data contain error and that can result in bad decisions. We will have to work hard to control the error in order to allow the data to inform us and help us make our decisions.

Facts. You may have noticed that we haven't defined the term, *fact*. This is not an accident. Statisticians rarely use the term in any technical sense. They consider it a philosopher's term.

You may have heard the expression, "It's a statistical fact!" but you probably didn't hear that from a statistician. The meaning of this expression is unclear. It could mean that a statistical description is free from error, which is never the case. It could mean that the results of a statistical inference are certain, which is never the case. It probably means that a statistical conclusion is good enough to base our decisions on, but statisticians prefer to state things more cautiously. As we mentioned earlier, statistics allows us to say how good our statistical conclusions are. Statisticians prefer to say how good, rather than just to say, "good enough."

Some philosophers say that facts are the things we can measure, even if we don't measure them. Judy is some height or other, even if we don't know what that height is. Other (smarter) philosophers say that facts are the results we would get if our measurements could be free of error, which they can never be. This sort of dispute seems to be an excellent reason to leave facts to the philosophers.

LEVELS OF MEASUREMENT

You may have noticed that we have cheated a bit. In Chapter 1 "Statistics for Business," we defined statistics as the use of numbers to describe general facts about the world. Now, we have shown how some measurements used in statistics are not really numbers at all, at least not in the ordinary sense that we learned about numbers in high school. Statistics uses an expanded notion of number that includes other sorts of symbol systems. The statistical notion of number does have its limits. First of all, the non-numeric values used in statistics must be part of a formal system that can be treated mathematically. In this section, we will learn about the most common systems used in statistics. Also, for most statistical techniques used in inferential statistics, the values will need to be converted into numbers, because inferential statistical techniques use algebra, which requires numbers.

Let's start with our example of measuring Judy's height. We say that that measurement results in a number, 62. You may remember from high school algebra (or else from Appendix A) that there is more than just one kind of number. There are counting numbers, integers, rational numbers, real numbers, and so forth. We will see that it matters a lot what kind of number we use for different kinds of measurements. Height is measured with positive real numbers. A person can be 5 foot $10\frac{1}{2}$ inches tall, but they can't be minus six feet tall, or zero inches tall.

We can see that the type of number used for different kinds of measurement depends on what different values are possible outcomes of that type of measurement. The number of items on a receipt is measured as a positive integer, also known as a counting number. Counting numbers are non-negative integers because counts don't include fractions (ordinarily) or negative values. The number of children in a family could be zero (technically, a non-negative integer). A bank balance, whether measured in dollars or in cents, is an integer, because it can be negative as well as positive (negative if there is an overdraft), but we can't have fractions of pennies. Height and weight are positive real numbers. The amount of oil in an oil tanker could be zero as well as a positive value. So it is measured as a non-negative real number. The temperature inside a refrigerated container could be negative or positive or zero, at least in the Celsius or Fahrenheit scales.

KEY POINT

In algebra, different types of numbers are defined in terms of the different possible values included. We choose the type of number for measuring a particular type of variable when the different possible numeric values match up to the different measurement outcomes.

But what about measurements that don't result in numbers? Let's go back to our example of making an inventory of cars in a car lot. Suppose that each parking spot in the car lot is labeled from A to Z. Each car is either a sedan, convertible, or minivan. Our inventory sheet, shown in Table 2-1, has one line for each parking spot on the lot. We go through the lot and write down the model of the car in the line corresponding to its parking spot.

Car models, like height, or weight, or dollars in a bank account, have different values for different subjects, but the different values don't really correspond well to the different values for different types of numbers. The closest match is positive integers, by assigning different numbers to different models, like 1 for sedan, 2 for convertible, and 3 for minivan, but there is a problem with this as well.

Table 2-1   Automobile inventory.

Parking spot	Type of car
A	sedan
B	sedan
C	convertible
D	sedan
E	minivan
F	minivan
…	…

Integers are different from car models in two ways. The first problem is minor. There are an infinite number of integers, but only a finite number of car models. Every bank account may have a finite amount of money in it, but in principle, there is no limit to how much money can be in our bank account. That is a good reason to use integers to measure money. Similarly, new car models, like the minivan, occasionally get invented, so the infinite number of integers available may be handy.

The other problem is not so minor. The integers possess a very important property that car models do not: the property of order. Three is bigger than two, which is bigger than one. There is no relation like "bigger than" that applies to car models. The best way to see this is to realize that there is no reason to choose any particular number for any particular car model. Instead of choosing 1 for sedan, 2 for convertible, and 3 for minivan, we could just as easily have chosen 1 for convertible, 2 for minivan, and 3 for sedan. Our choice of which number to use is *arbitrary*. And arbitrary is not a good thing when it comes to mathematics.

Statisticians do not classify different types of measurement in terms of what types of numbers (or non-numerical symbols) are used to record the results. While it may make a difference to certain types of calculations used in statistics as to whether the original measurements are integers or real numbers, this difference does not figure into the classification of measurement. Instead, they group the different types of numbers in terms of what makes a difference in using different statistical techniques. Just as with statistical assumptions, the different types of measurement, called *levels of measurement*, are grounded in the very important issue of how to pick the right sort of statistical analysis for the problem at hand. The different levels of measurement are:

- *Nominal scale*. When the values have no relation of order, the variable is said to be on a nominal scale. This corresponds to categorical data. Example: Methods of drug administration: oral, intravenous, intramuscular, subcutaneous, inhalant, topical, etc.
- *Ordinal scale*. When the values have a relation of order, but intervals between adjacent values are not equal, the variable is said to be on an ordinal scale. This is one type of numerical data. Example: Coin grades: Poor, Fair, Good, Very Good, Fine, Very Fine, Extra Fine, Mint, etc.
- *Interval scale*. When the values have a relation of order, and intervals between adjacent values are equal, but a value of zero is arbitrary, the variable is said to be on an interval scale. This is another type of numerical data. Example: Fahrenheit temperature.
- *Ratio scale*. When the values have a relation of order, the intervals between adjacent values are equal, and a value of zero is meaningful, the variable is said to be on a ratio scale. (A meaningful value of zero is called a *true zero point* or *origin*.) This is the last type of numerical data. Example: Money, with debt measured as negative numbers.

HANDY HINTS

Some textbooks define ordinal data as a form of categorical data and others as a form of numerical data. This is because ordinal data has characteristics of each and, depending on what we do with it, it may be treated as either. An ordinal variable does classify each individual subject item into one and only one category and, by that standard, is definitely a type of categorical variable, where the categories have a specific order. When graphing, ordinal variables are treated as categorical. Because the positive integers are a very convenient way of showing order (after all, we are all pretty familiar with the counting order), ordinal variables are very often coded numerically as positive integers, which is one reason why some textbooks classify ordinal variables as numerical.

Finally, many statistical inference techniques that require an interval level of measurement can be and are used effectively with ordinal variables coded as integers. (This is a good example of using a statistical technique even though one of its *assumptions* is violated.) When it comes to inferential statistics, ordinal variables are treated as categorical or numerical depending on the technique used. Using a technique (called a nonparametric technique) designed for categorical variables will be more accurate, but may be less powerful. (That is, the technique is more likely to fail to give a definitive answer to our question.) Using a technique (called a parametric technique) designed for numerical variables is more powerful, but less accurate, because the fact that the adjacent categories of an ordinal variable are not guaranteed to be equally far apart violates one of the assumptions of the technique.

There is also a special case of a nominal variable that can be treated as interval. When a variable can take on only two values, like true and false, or male and female, or is-a-current-customer and is-not-a-current-customer, the data are nominal because there is no order to the values. When used in inferential statistics, these variables can be treated as interval, because, having only two possible values, they only have one interval between the values. And one interval is always equal to itself. Variables that can take on only two values are sometimes called *binary variables*, most often called *dichotomous variables*, and when used in the inferential technique known as *regression* (see Chapter 12 "Correlation and Regression"), as *dummy variables*.

We will learn more about all of this in Part Three, where we learn about inferential statistical techniques.

Note that this classification system ignores the differences between integers, rational numbers, and real numbers. This is because measurements are always made up to some level of precision. There is always the possibility that two values are so close that they cannot be distinguished. Two people, where one is six feet tall and the other is six feet and one millionth of an inch tall, will both be classified as six feet tall. For the purpose of the analysis, there is no difference between them. There are no truly continuous numbers in measurement. Since statistics always begins with measurement, the issue of continuity is irrelevant in applied statistics.

The only exception to this rule is for measurements that don't ever come in fractions. For example, sometimes the general fact of the world we care about is discovered by counting, as in the number of widgets we produced last week. The number of widgets is always a whole number. It wouldn't make much sense to say we have $45\frac{1}{2}$ widgets on hand. As we will see in later chapters, statistics handles this problem in two different ways. If the number of items is large enough, many of our questions can be answered statistically by pretending that fractional values are possible. For example, if we are producing between 40 and 50 thousand widgets a month, the fact that the detailed calculations use fictitious values like 42,893.087 instead of genuinely possible values like 42,893, doesn't matter much. If the number of items is small (usually less than 20), and it is the count that we really care about, there are separate statistics, called *count statistics* that are used to answer our questions. In order to keep this difference straight, we will have two separate examples running through the book: one about counting sheep, and one about measuring people.

As we will see later on in Part Two and Part Three, the issues of possessing order, equal intervals, and a true zero point are used to classify variables because they make a difference as to whether different statistical measures and techniques can be used effectively.

Error

In order to help make decisions, we need to know the true value of the information that statistics provides. Statistics not only provides information, but also specific measures of the degree of confidence with which that information can be trusted. This ability to measure the quality of statistical information is based on the concept of error.

TIPS ON TERMS

Error. The degree to which a description does not match whatever is being described.

All aspects of statistics are prone to error. No individual measurement is free from error. Measurement is a human process, limited by our tools and our senses and our other fallible human capacities. We need to understand measurement error in order to have the right amount of confidence in our data. Statistical measures and statistical techniques are also prone to error of another type. Even when calculated mechanically and exactly from the data, the information statistics gives us is never an exact description of the true state of the world. (We will see more of why this is so later on in this chapter and also in Chapter 3 "What Is Probability?") The statistical theory of error helps us gauge the right amount of confidence to have in both our data and our statistics.

CLEANING YOUR DATA

Computers have made statistics much easier to do, but they also make it much easier to do statistics badly. A very common and very bad mistake is to collect our data, get it onto the computer, and immediately begin to calculate statistics. Both during and immediately after collecting data, we must check our data thoroughly for errors. We will not be able to find every error. There are many types of errors we can't even find in principle. But when a value is clearly wrong, we need to fix it, or throw it out.

Throwing out a value leaves what is called *missing data*. Missing data can be a real problem in statistics, but missing data is better than wrong data.

CRITICAL CAUTION

Missing Data
When there are multiple variables for each subject and one or more values for a subject are missing, various serious problems can occur with different statistical measures and techniques. Most computer programs that do statistics will handle missing data automatically in the simplest way possible, which is usually good enough. However, when there is lots of missing data, an expert should be consulted to determine the best way to treat it.

QUICK QUOTE

There is only one good way to deal with missing data. Don't have any!
Gertrude Cox

How do we know when the data are bad? Often, it's quite simple. If the variable is age, then values like "handle," "−3," and "123" are most likely errors. Before data are collected, it is important to determine what the acceptable values will be. These acceptable values are called *legal values*. When the variable is non-numerical, it is a good idea to set up specific values called *codes* for each legal category. Returning to our example of car models, we might decide to save time and trouble by just coding the model of each car using the first letter: *C* for convertible, *S* for sedan, and *M* for minivan. This is fine, unless we find a coupe on the lot! Always plan for your possible values before you start collecting data. If you are not sure of all possible values, have a system ready to add more legal values and validate them.

BIO BITES

Always Almost Always
There are also more indirect ways of finding bad data. The first author used a multiple-choice questionnaire for his Master's research. All of the items had answers rated from "1" to "5," ranging from "never," through "sometimes" to "always."

The answers for one subject were all "4." Either the computer was broken that day, or that student was in a hurry and didn't want to read the questions.

You should also consider how the data will be collected. For instance, if we are collecting information about cars in the lot on handwritten sheets, different sorts of errors are likely to occur than if we are collecting that same information with a hand-held computer. We should plan our codes accordingly. If we are using the computer, we may want to use the full names of the colors of the cars in the lot. If we know all the colors in advance, we could put them on a menu. If we are writing things down by hand, names can be a problem. The word "gray" can look an awful like the word, "green." It might be better to assign numbers for each color and list those numbers at the top of the sheet.

The important lesson is that dealing with error starts even before data is collected. Careful planning and design is needed to prevent errors from happening to begin with, and to make errors easier to detect if they do happen. We cannot prevent errors entirely, but we need to work carefully to minimize them.

TWO WAYS OF BEING WRONG: VALIDITY AND RELIABILITY

In later chapters, we will have a great deal more to say about error. For now, it is important to understand that there are two sorts of error. In statistics, these two kinds of error are talked about in terms of *reliability* and *validity*. The distinction is related to the difference between precision and accuracy in physics and engineering, or between precision and clarity in philosophy.

Suppose I am shooting at a target with a bow and arrow. Over time, I find that I am hitting the target about 30% of the time, but that almost all of my misses are falling short of the target. In addition, my arrows are scattered up and down, right and left. The first step is to realize that I am making two errors. My precision is low—the arrows are going all over the map. And my accuracy is low—I am hitting consistently short of the target.

Being a statistician—and perhaps not a good student of archery—I choose to work on my precision first. I give up on trying to hit the target, and I just try to get all of my arrows to land in a small area, well short of the target. Once I have accomplished this, I am making just about the same error with every shot—I am always in line to the target, and I am always falling short.

My precision is high—I hit almost the same spot every time. My accuracy is low—I never hit the target.

At this point, I go to an archery instructor. I say, "I've gotten very good at getting all the arrows to land in the same spot. But I'm pulling the bow as hard as I can, and they don't go far enough." He says, "Let me watch." I shoot ten arrows. They all land in the dirt short of the target, in a circle smaller than the bull's eye of the target. He laughs, "You don't need to pull any harder. A bow should always be pulled with just enough strength for the arrowhead to be just past the bow. If you want to hit the target, you have to shoot farther. To shoot farther, just aim higher." I give it a try, and, with a little practice, I am hitting the target dead center every time. I've corrected my second error. I'm shooting accurately.

When we are both precise and accurate, we hit the target. In statistics, we would say that when our measurements are both reliable and valid, we have reduced both types of error.

HANDY HINTS

Reliability is like precision and validity is like accuracy.

A similar situation happens in golf. If my shots consistently go left, the golf pro coaches me to improve parts of my swing to reduce hooking. Likewise for going right and slicing. The coach is working to reduce the *bias* in my form and my golf swing. None of the coaching will have anything to do with aiming at the target. It will all have to do with my form. On the other hand, if I am missing both left and right, the golf pro will assist me with my aim, that is, keeping my shots on target, keeping the spread down. The golf pro is working first to reduce bias, that is, to increase accuracy, so that my shots are centered around the hole. Secondly, the pro will help me increase the precision of my golf shot, so I'm not just getting somewhere near the hole, I'm landing on the green, very near the hole.

For reasons we will see in a moment, in statistics, we have to do things in the reverse order from what our golf pro did and from what is done in sports in general. First, we need to get the spread down, increasing the reliability of our measurements, and then we need to make sure we are pointed in the right direction, increasing their validity. (This is how our statistician tried to teach himself archery, and why the archery instructor found it so amusing.)

Reliability is how statisticians talk about minimizing unbiased error, reducing spread. The value of knowing the reliability of our measurement is that we don't have to measure again and again to get it right. If our technique for measuring Judy's height is reliable, whatever height we get the first time won't be too far from the height we get the second time or the fifth time or the fiftieth time (presuming Judy's real height isn't changing between measurements). We can *rely* on the number we get being independent of when we measure it. Measuring a person's height with a tape measure is pretty reliable; that is, if we measure several times in a row, we'll probably get almost the same answer.

Validity is how statisticians talk about minimizing biased error, making sure things are centered at what they are pointed at. The value of knowing the validity of our measurement is that we have a good estimate of how far-off from the truth our measurement can be. If our technique for measuring Judy's height is valid, we know that her real height won't be far from what we get by measuring. If our measuring technique is not valid, we will need to find and correct the source of bias if we can, or take it into account and adjust for it. For example, if our tape measure is cut off at the front end, and starts at two inches, instead of at zero, every time we measure Judy's height, our result is two inches taller than her actual height. Getting a good tape measure would eliminate the bias.

There is an interesting relationship between reliability and validity. Our measurements can't be any more valid than they are reliable. The amount of reliability is a ceiling on the amount of validity. This is true with precision and accuracy as well. We can be perfectly precise and very inaccurate. In golf, if I hook badly, it doesn't matter if my spread is perfect. If it was, I might find myself always missing the hole to the left by exactly ten and a half feet. Strange, but possible. A statistician would say that my shot was biased and invalid, but highly reliable. A physicist would say that my golf shot was inaccurate, but precise. And my golf coach would tell me to pretend that the hole was ten and a half feet further to the right.

On the other hand, we can't be perfectly accurate and very imprecise. If my every shot is a hole in one, then the spread of all my shots can't be wider than the hole. To have a high degree of accuracy, we need to have both validity and reliability; we need to be both free of bias and consistently close to the target. And, if our reliability is low, then we can't know for sure whether our validity is good. If we may always be missing by ten feet or so, we can't find a bias of less than ten feet with any certainty.

Another way to think about this is in terms of a clock. If our clock runs with perfect precision and we set it five minutes fast, it will never give us the correct time, but it will always be exact. It will be exactly five minutes off. On the other hand, if the clock has poor precision, running faster and slower from time to time due to a broken regulator, it will only give the correct time now and then, and we won't have any way of knowing when it is right. We will also have no way to set it to the correct time and keep it there, because it does not keep time reliably.

In statistics, there is an important difference between reliability and validity. We can calculate the reliability without even knowing the right answer! Let's go back to the golf example. Suppose I take a bunch of shots at a hole from a place where I can reach the green easily. Now, we go up in the blimp and take a picture of all of the golf balls from straight overhead. Suppose we can see the golf balls in the picture, but we can't see the hole, because someone removed the flag. If all of the golf balls are close together, we will know that my shooting was very precise, very reliable, but we won't know if I was hooking or slicing or very accurate. Now, someone goes and puts the flag back in the hole, and the cameraman takes another photo. If the hole is near the center of the area where the balls were found, then my golf shot was accurate, free of bias, or, in statistical terms, valid. We need to see the target to determine accuracy. In assessing validity, like accuracy, we need to know what the true value is.

When it comes to statistics, obviously, validity is the most important thing. We want our numbers to be right, or at least clustered around the right answer. But validity is much harder to measure than reliability. The reason for this is that we don't know the world directly; we only find out about the world by observing it. Recall that measurement is just formalized, repeatable observation. As a result, we are always comparing one observation to other observations, one measurement to other measurements. Statistics is like playing golf, only nobody knows exactly where the hole is. Suppose we measure Judy's height over and over again and record the numbers. If all of the numbers are close together, we know that our technique for measuring Judy's height is reliable, but how do we know if it is valid? Maybe, like the case with the cut-off tape measure, every measurement is almost exactly two inches off. Unlike the golf balls on the golf course, there is no way of knowing where the target is. What is Judy's "true height"? The only way we know Judy's height at all is to measure it, yet we don't know if our measuring technique is giving us the right answer.

BIO BITES

Counting Blood Cells

The co-author of this book worked at a hospital blood lab when he was in high school. A new machine for counting red blood cells had just been invented. It gave different results than the old machine. Was it broken? Possibly not. Maybe it was better than the old machine. If the old machine had a bias, and the new one didn't, then the more accurate results would simply look different—they would look wrong from the perspective of the old way of doing things.

This is the difficulty of determining validity. Only if we know what is really out there can we say which method of measurement is more valid. But the only way to know what is out there is to measure it, one way or another.

The hospital tested the new machine by comparing it against two or three other methods, and determined it was a better device than the one it was replacing.

The best way to determine validity is to compare the measurements we get to other measurements taken using an entirely different measurement technique. We could compare our results measuring Judy's height with other measurements taken with a doctor's scale. When there is only one way to measure something, the problems of assessing validity become much more difficult.

Because of these two facts about the relationship between reliability and validity, in statistics, we always consider reliability first. First of all, reliability is easier to measure, because we don't have to know where the target is. This is the opposite of archery and golf, where we can see the target, and so the easiest thing is to evaluate each shot with respect to the target. Even more importantly, because our measurements can be no more valid than they are reliable, it makes no sense to attempt to check our validity if our measurements are all over the place. As we said above, low reliability means we can't even measure validity very closely. If all our golf shots are flying into the crowd, it really doesn't matter if more of them are going to the right than to the left.

Sampling

We said earlier that, even if our measurements were perfectly free from error, statistics would still not give us perfectly correct answers. Over and above measurement error, there is also statistical error. Key to understanding statistical error is the concept of sampling. Sampling is the process by which we choose the individuals we will measure. The statistical errors due to limitations of sampling are known as *sampling error*.

Statistical conclusions, whether the results of measurements or the results of an analysis, usually take the form of a single number (the statistic, which is a general description) that characterizes a group of numbers (the data, which are specific descriptions). But we may want to know a general fact about subjects we cannot measure. A good example of this is political polls for predicting election results. Before the election, the pollsters call people up and ask who they are going to vote for. Even if we supposed that everyone knows who they will vote for, that everyone answers, and that everyone tells the truth (all of which means that there is no measurement error), the pollsters could make the wrong prediction. Why? Because there is no way the pollsters can call every voter. We all see polls on TV when no one called us the night before. They must have been calling someone else. Suppose the pollsters only called Republicans that night? Their prediction might be way off.

WHAT IS A POPULATION?

Ideally, if the pollster could call every person who was going to vote (and there was no measurement error), they could get an exact prediction of the election results. The group of people who are actually going to vote in the election is what statisticians call the *population*. Practically speaking, limits on time and money usually prevent measuring values from the entire population, in polls or elsewhere. However, there are problems measuring the entire population, even in principle. Even the night before the election, some people might not be sure if they are going to vote. Maybe they are running late the next morning and decide to skip it. Then, at lunch, a few co-workers decide to go to vote together and the person who missed voting that morning gets dragged along. Even someone who is 1000% sure they are going to vote tomorrow may have an emergency and just not be able to make it. And we also have to consider someone who plans to vote, does vote, and whose ballot gets eliminated later on due to damage from a broken voting machine.

CRITICAL CAUTION

A population is a *theoretical* concept. We can envision it, but, when we get down to the nitty-gritty details, we can almost never actually measure it exactly.

It is easy, but wrong, to think of a population as something real, that we can't measure because of the expense, but there are always limitations. Some of these limitations might be classified as measurement error, and others might not, but the result is the same. Suppose we want to evaluate yesterday's sales. Then yesterday's sales are our population. Yesterday's sales receipts are how we can measure them. It may look like we have access to the entire population at low cost, but that is not the case. Yesterday's sales are past events. Absent a time machine, we will never see them again directly. The

sales receipts are just records, measurements of those events. Some may be lost or have errors. Or a sales receipt from some other day may be marked with yesterday's date by mistake.

KEY POINT

The most important thing to understand about populations is the need to specify them clearly and precisely. As we will see later on, every statistical study begins with a question about some population. To make that question clear means being clear about what the population is, who or what is or is not subject of the study, what is the study question about. A good statistical study begins with a clearly specified question. The easiest way to turn a good study bad is not to specify the population of interest clearly and precisely.

In fact, one of the key reasons that different pollsters and pundits had different predictions for the results of the Iowa Democratic Caucus is that they had different expectations about who would participate in the caucus, that is, who would be in the population.

The example of yesterday's sales receipts is the ideal situation. Absent measurement error, we have every reason to believe that we have access to the entire population. Our collection of receipts is what statisticians call a *comprehensive sample*. This is one of the best types of sample to have, but, in practice, it is usually impossible to get. And, when we have it, it may be too costly to measure every individual in the sample.

TIPS ON TERMS

Population. All of the subjects of interest. The population can be a group of business transactions, companies, customers, anything we can measure and want to know about. The details of which subjects are and are not part of our population should be carefully specified.
Sample. The subjects in the population we actually measure. There are many ways of picking a sample from a population. Each way has its limitations and difficulties. It is important to know what kind of sample we are using.
Sampling. The process of selecting the individuals from the population that makes up our sample. The details of the sampling procedure arc what make for different kinds of sample.

WHAT IS A SAMPLE?

This brings us to the critical notion of a *sample*. A sample is the part of the population we actually measure. *Sampling* is the process of selecting those members of the population we will measure. Different ways of sampling lead to different types of samples. The types of statistical error we can encounter in our study depend on how our sample differs from the population we are interested in. Understanding the limits of how confident we can be about the results of our study is critically tied to the types of statistical error created. Choosing the right sampling procedure and knowing the errors it creates is critical to the design and execution of any statistical study.

KEY POINT

Choosing the right sampling procedure and knowing the errors it creates is critical to the design and execution of any statistical study.

The relationship between sampling and error is not as hard as it seems. We begin by wanting to know general facts about a situation: What were last year's sales like? How will our current customers react to a price increase? Which job applicants will make the best employees? How many rejects will result from a new manufacturing process? If we can measure all of last year's sales, all of our current customers, all of our future job applicants, etc., we will have a comprehensive sample and we will only have to worry about measurement error. But to the degree that our sample does not include someone or something in the population, any statistics we calculate will have errors. General descriptions of some of last year's sales, some of our current customers, or just the current crop of job applicants will be different from general descriptions of all of the sales, customers, or applicants, respectively. Which members of the population get left out of our measurements determine what the error will be.

HANDY HINTS

Note that sampling error is a question of validity, not reliability. That is, sampling error introduces bias. Differences between the sample and the population will create statistical results that are different from what the results would have been for the entire population, which is what we started out wanting to know. On the other hand, our choice of sample size affects reliability. The larger the sample size in proportion to the population, the more reliable the statistics will be, whether they are biased or not.

Here are some of the most common types of samples:

- *Comprehensive sample.* This is when the sample consists of the entire population, at least in principle. Most often, this kind of sample is not possible and when it is possible, it is rarely practical.
- *Random sample.* This is when the sample is selected randomly from the population. In this context, *randomly* means that every member of the population has *an equal chance of being selected* as part of the sample. In most situations, this is the best kind of sample to use.
- *Convenience sample.* This is usually the worst kind of sample to use, but, as its name implies, it is also the easiest. Convenience sampling means selecting the sample by the easiest and/or least costly method available. Whatever kinds of sampling error happen, happen. Convenience sampling is used very often, especially in small studies. The most important thing to understand about using a convenience sample is to understand the types of errors most likely to happen, given the particular sampling procedure used and the particular population being sampled. Each convenience sampling process is unique and the types of sampling error created need to be understood and stated clearly in the statistical report.
- *Systematic sample.* This is when the sample is selected by a non-random procedure, such as picking every tenth product unit off of the assembly line for testing or every 50th customer off of a mailing list. The trick to systematic sampling is that, if the list of items is ordered in a way that is unrelated to the statistical questions of interest, a systematic sample can be just as good as, or even better than, a random sample. For example, if the customers are listed alphabetically by last name, it may be that every customer *of a particular type* will have an equal chance of being selected, even if not every customer has a chance of being selected. The problem is that it is not often easy to determine whether the order really is unrelated to what we want to know. If the stamping machine produces product molds in batches of ten, choosing every tenth item may miss defects in some part of the stamping mold.
- *Stratified sample.* Also called a stratified random sample. This is a sophisticated technique used when there are possible problems with ordinary random sampling, most often due to small sample size.

It uses known facts about the population to systematically select subpopulations and then random sampling is used within each subpopulation. Stratified sampling requires an expert to plan and execute it.

- *Quota sample.* This is a variant on the convenience sample common in surveys. Each person responsible for data collection is assigned a quota and then uses convenience sampling, sometimes with restrictions. An advantage of quota sampling is that different data collectors may find different collection methods convenient. This can prevent the bias created by using just one convenient sampling method. The biggest problem with a quota sample is that a lot of folks find the same things convenient. In general, the problems of convenience samples apply to quota samples.

- *Self-selected sample.* This is a form of convenience sample where the subjects determine whether or not to be part of the sample. There are degrees of self-selection and, in general, the more self-selection the more problems and potential bias. Any sampling procedure that is voluntary for the subjects is contaminated with some degree of self-selection. (Sampling invoices from a file or products from an assembly line involves no self-selection because invoices and products lack the ability to refuse to be measured.) One of the most drastic forms of self-selection is used in the Internet polls common to TV news shows. Everyone is invited to log onto the Web and vote for this or that. But the choice to view the show is self-selection, and others do not get the invitation. Not everyone who gets the invitation has Internet access. Since having Internet access is a personal choice, there is self-selection there, as well. And lots and lots of folks with Internet access don't vote on that particular question. The people who make choices that lead to hearing the invitation, being able to vote, and voting, are self-selected in at least these three different ways. On TV, we are told these polls are "not scientific." That is polite. Self-selection tends to create very dangerous and misleading bias and should be minimized whenever possible.

We will have much more to say about exactly what kinds of errors result from sampling in Chapters 3, 8, and 11. There is always more to learn about sampling. Note that, although we discussed measurement first, the practical order is: Define the population; Select the sample; Take the measurements. When we have that, we have our data. Once we clean up our data—see Chapter 6 "Getting the Data" about that—we are ready to analyze the data.

Analysis

Analysis is the process that follows measurement. In Chapter 1 "Statistics for Business," we discussed the difference between descriptive and inferential statistics. Both of these are types of statistical analysis. Here, we will explain those differences in more detail.

Our data consist of a number of measurements of one or more different features for each one of all of the individual subjects in our sample. Each measurement value gives us specific information about the world. We use mathematics to calculate statistical measures from those measurement values. Each statistical measure gives us general information about the world because it is calculated from multiple data values containing specific information. The process of calculating general information from data is called *statistical analysis*.

SUMMARIZING DATA: WHEN IS A NUMBER A STATISTIC?

Within the field of statistics, the word, "statistic" has another, more specific meaning. A statistic, also called a *statistical measure*, is a value calculated from more than one data value, using a specific calculation procedure, called a *statistical technique* or *statistical method*. We have mentioned one statistic, the *count*, in Chapter 1. We will learn about a number of other statistical measures in Chapter 8 "Common Statistical Measures." Examples of statistical measures are: ratio, mean, median, mode, range, variance, standard deviation, and many others.

STUDY REVIEW

In statistics, a statistical measure, is a variable calculated from the data. We discuss the most basic of these in Parts One and Two, especially in Chapter 8. Each variable is calculated using a specific method, described by a mathematical equation.

A *statistical procedure,* some of which are called *statistical significance tests*, are more complex methods that give you more advanced statistical measures. We discuss these in Part Three. Statistical procedures often involve a number of equations and provide more subtle and intricate information about the data. However, there is no hard and fast rule dividing the measures from the procedures. In all cases, a number is calculated from the data that informs us about the data.

The procedures used for calculating a statistical measure starts with multiple values and *summarizes* them, producing a single number that characterizes all of the values used in the calculation. It is this process of summarization that generates general descriptions from specific ones.

As we discussed in Chapter 1, there are two basic kinds of statistical measures, descriptive and inferential. As you might imagine, a descriptive statistic is one that describes a general feature of the data. An inferential statistic describes the strength of a relationship within the data, but its most common use is to say whether a relationship in the data is strong enough to affect the outcome of a particular sort of decision. The calculated value of the inferential statistic determines the conclusion of the statistical inference. For example, in one of the most basic inferential procedures, the *t* test, the end result is the calculation of an inferential statistical measure called the *t* statistic. The *t* statistic is higher whenever the value of the particular variable being analyzed is higher for one group of subject units than for another.

KEY POINT

Both descriptive and inferential statistics tell us about the world. An inferential statistic also answers a specific type of question within the framework of a statistical technique designed to perform a statistical inference. (For more on statistical inference, see the sidebar on inductive inference.) All of the guarantees for that statistical technique come with the proper use of the inferential statistic.

In the end, the distinction between a descriptive and an inferential statistic is not a hard and fast one. It is a common error in statistics to use a descriptive measure as if it could provide a conclusion to a statistical inference. It is a common oversight in statistics to forget that any inferential statistic does describe the data in some way. Simply put, every inferential statistic is descriptive, but most descriptive statistics are not inferential.

WHAT IS A STATISTICAL TECHNIQUE?

Throughout these first two chapters, we have talked about statistical techniques and differentiated them from statistical measures, but we haven't yet defined the difference. A statistical measure is a number that results from making calculations according to a specified procedure. For every statistical measure, there are one or more (usually more) procedures that produce the right number as a result. Take the example of the simplest statistical measure, the count. The procedure used to produce counts is counting, which we all know how to do.

When we get to more sophisticated statistical measures, particularly inferential statistical measures, the procedures for calculating the measure get a lot more complex. We call these much more complex procedures *statistical techniques* or *statistical methods*. As a result, the distinction between a simple calculation procedure and a complex statistical technique is also not a hard and fast one.

One way of teaching basic, descriptive statistical measures is to present step-by-step procedures for calculating them. On the other hand, this method is almost never used for the more complex inferential measures, except in the most advanced texts. Knowing how to do these calculations may be a good teaching device, but, on the job, no one does these calculations, even the simplest ones, by hand anymore. Computers are used instead. In this book, we will not walk through the details for calculating most statistical measures, because those can be found in other excellent texts, which we list for you at www.qualitytechnology.com/books/bsd.htm. We will, however, provide detailed procedures for some special types of calculations that you may find useful in business when there is no computer around. (Recall the quotation from John Tukey in the introduction to Part One about the stick in the sand on the beach. Even without a computer, we can learn important facts about data right on the spot.)

FUN FACTS

Brewing Up Inferential Statistics

Until the early part of the last century, statistics was about description. Then, in 1920, a statistician named Gossett, working in the business of brewing beer for Guinness, came up with a trick called the *t* test. A famous statistician and scientist named Sir Ronald A. Fisher immediately recognized the enormous importance of the *t* test, and began the development of a second kind of statistics, inferential statistics.

Statistical methods are formal, which means that once we abstract away from the topic of interest by measuring things, we can do statistics on almost anything: employees, receipts, competitors, transactions, etc. But the guarantee that statistical techniques provide is not apodictic, because of the possibility of statistical error. As we discussed before, even if all our measurements are perfect, our conclusions are not guaranteed to be true.

What Fisher recognized was that the *t* test (also called Student's *t* test, because Gossett had to publish under the pseudonym "Student," because, at the time, Guinness Breweries prohibited its employees from publishing their work in scholarly journals) provided a weaker sort of guarantee, based on the concept of probability. If all of our measurements are perfect (that is, all of our premises are true), we have a guarantee that the statistical values we calculate are probably close to the right values. (We will learn more details about this guarantee in later chapters.)

The three most important things to understand about statistical inference are that it uses a specifiable *procedure*, that procedure is *formal*, and that it uses *probability* to describe how confident we have a right to be about the results. Today, formal procedures can be performed by computer, which is what makes the very powerful and very complex statistical analyses so popular and useful in business (and elsewhere) possible.

Quiz

1. What is the correct order of the first three steps in performing statistics?
   - (a) Analysis, sampling, and measurement
   - (b) Sampling, measurement, and analysis
   - (c) Analysis, measurement, and sampling
   - (d) Measurement, sampling, and analysis

2. Which of the following statements about measurement is not true?
   - (a) Measurement is a formalized version of observation
   - (b) Measurement is different from ordinary observation
   - (c) Measurement provides a specific description of the world
   - (d) Measurement provides a general description of the world

3. How is a variable used in statistics?
   - (a) A variable usually corresponds to some measurable feature of the subject
   - (b) A variable is a person, object, or event to which a measurement can be applied
   - (c) A variable is the result of a particular measurement
   - (d) A variable is the collection of values resulting from a group of measurements

4. The series "President, Vice-President, Secretary, Treasurer, Board Member" is on which type of scale?
   - (a) Nominal
   - (b) Ordinal
   - (c) Interval
   - (d) Ratio

5. Which of the following components of statistics contain error?
    (a)   Measurement
    (b)   Statistical analysis
    (c)   Sampling
    (d)   All of the above

6. If we have a set of measurements that are valid, but not very reliable, they will...
    (a)   Be clustered around the right value, but in a wide cluster
    (b)   Be clustered very closely together, but around the wrong answer
    (c)   Be in a wide cluster around the wrong value
    (d)   Include at least one measurement that is exactly the right value

7. Validity is how statisticians talk about minimizing _____ error; Reliability is how statisticians talk about minimizing _____ error.
    (a)   Biased; biased
    (b)   Biased; unbiased
    (c)   Unbiased; biased
    (d)   Unbiased; unbiased

8. When a comprehensive sample is not possible, what is the best sampling technique to use in order to avoid introducing additional bias?
    (a)   Convenience sample
    (b)   Stratified sample
    (c)   Random sample
    (d)   Systematic sample

9. Which of the following is the end product of the procedures used for calculating a statistical measure?
    (a)   A single summary number that characterizes all of the values used in the calculation
    (b)   A statistical technique
    (c)   A range of numbers that characterize the population of interest
    (d)   A valid and reliable measure

10. Every inferential statistic is _____, but most descriptive statistics are not _____.
    (a)   Inferential; inferential
    (b)   Inferential; descriptive
    (c)   Descriptive; inferential
    (d)   Descriptive; descriptive

CHAPTER

3

What Is Probability?

Probability has an important role in statistical theory. Its role in learning about statistics is less clear. However, many statistical texts cover basic probability and readers of this book who want to do well in their statistics class will need to understand probability, because it will probably be in the exam. Here, we use the notion of probability to introduce the important statistical notions of independence and distributions, which will come up again throughout the book.

READING RULES

This is the first chapter in which we will be using mathematics. There will be some equations, which, if you are taking a course, you may need to memorize for exams. Here, we will focus on explaining them. By the last few sections of the chapter, we will be ready to do our first real statistics. But, even for that, there will be almost no math required.

How Probability Fits in With Statistics

Thomsett (1990) points out that, in some ways, probability and statistics are opposites. Statistics tells us general information about the world, even if we don't understand the processes that made it happen. Probability is a way of calculating facts about the world, but only if we understand the underlying process. One way that probability fits in with statistics is that, in order to prove that this or that statistical technique will actually do what it is supposed to do, statistical theoreticians make assumptions as to how the underlying process works and use probability theory to prove mathematically that statistics will give the right answer. Obviously, to us, as users of statistics, that kind of theoretical connection between probability and statistics isn't too useful, although knowing that it is true can give us confidence that statistics actually works.

For us, the most important way that probability fits in with statistics is that it shows us the way that numbers calculated from a sample relate to the numbers for the population. Every element of statistics that we actually use in business or elsewhere is calculated from the sample, because the population, as we noted in Chapter 2 "What Is Statistics?" is just a theoretical abstraction. For every practical element of statistics based on the sample, there is a corresponding theoretical element based on the population. Once we understand the notion of probability, we will be able to see how the numbers we calculate from the sample can tell us about the real values—the true values, if we are willing to use that term—of the population that we would like to have, in the ideal, to help make our business decisions.

Measuring Likelihoods

Probability is the mathematical study of chance. In order to study chance mathematically, we will need some mathematical measure (not necessarily a statistical measure) of chance. The mathematical measure of chance is called *the probability of an event* and it is symbolized as $Pr(x)$, where x is the event. Probabilities are measured using a statistical measure called the *proportion,* symbolized as p. Probabilities are based on the notion of likelihood. In this section, we will explain the basics of chance, likelihoods, and proportions.

LIKELIHOODS AND ODDS: THE MYSTERIES OF CHANCE

What do we mean by *chance*? By *chance*, we mean the events for which we do not know the cause. Even if we believe that every event has a cause, often something will happen for no reason that we know of. Sometimes we say that this happened "by chance." Suppose that we walk into a store and the person just in front of us is awarded a prize for being the store's millionth customer. We might say that that particular customer turned out to be the millionth customer "by chance" even though, presumably, there were reasons why it was them (and not us).

Maybe we went back to check to see that the car door was locked, which delayed us by half a minute. Maybe they spotted a better parking spot that we missed, which put them closer to the door. Maybe we dropped by the store very briefly the night before because we couldn't wait until today for that candy bar. Had we stayed home hungry, they would have been the 999,999th customer today, and we would have been the millionth. When there is no clear line of simple causes, we use the word "chance."

Ignoring causes: talking about likelihood

The trick to probability is that, whether chance is about causes we don't know about, causes that are too complicated, or events that actually have no cause, we can talk about these events without talking about their causes. In ordinary day-to-day matters, we do this using the notion of *likelihood*. Some things are more likely to happen than others. Bob is usually late to meetings, so it is likely he will be late to this next meeting. Rush hour usually starts early on Friday, so, it is unlikely our delivery truck will have an easy time this Friday afternoon. The likelihood of our winning the lottery tomorrow is very low. The likelihood that the sun will rise tomorrow is very, very high. Whether we believe modern science, and think the Earth rotates, or we use the ancient Ptolemaic model that the sun circles the earth, doesn't matter. Our experience tells us that sunrise is a very likely event, independent of theory. Note that even though likelihoods may be due to many things, we often believe that the likelihood of something is high or low based on how often similar things have happened in the past. This is another case of the basic assumption that the future will be like the past that we mentioned in Chapter 1 "Statistics for Business."

Simple numbers for chances: odds

Even in ordinary day-to-day dealings, we deal with likelihoods in terms of numbers. One way we do this is with the notion of *odds*. When the likelihood is low, we say that the odds are against it. We also use odds to express likelihoods more exactly, with numbers. The odds of heads on the flip of a fair coin is 50–50. The odds on rolling a six on a single die is one-to-five (or five-to-one against) etc. Odds are based on a statistic called the *ratio*, which in turn is based on the statistic we learned about in Chapter 1, the *count*. Likelihoods cannot always be calculated using counts alone, but when they can be, then we use the notion of odds.

KEY POINT

Different situations lead to different ways that likelihoods can be calculated mathematically. This is very important in the philosophy of probability, although, as it turns out, not so much in the mathematical theory. As we will see later in this chapter, there are three different sorts of situations, leading to three different types of probability. In the first type of situation, we can count everything, which allows us to calculate the odds. In the second, we can use the past to estimate the likelihoods. In the third type, we can only use our own subjective intuition to guess at the likelihoods. In all three cases, the mathematical theory of probability works out the same (which is pretty remarkable, all things considered).

Let's return to our example of counting sheep from Chapter 1, to see what a ratio is and how it relates to odds. Suppose we have a small flock of sheep. We count the sheep and discover we have 12 sheep. Some of our sheep are black and others are white. Since the two colors of wool are sold separately for different prices, from a business perspective, the color of the sheep is important. The categorical (nominal) variable, *color*, may be *relevant* to our future business decisions as shepherds. Being smart businessmen, willing to use probability and statistics, we choose to measure it.

Categorical variables are measured by counting. We can count black sheep the same way we count sheep. Suppose we count the black sheep in our flock and discover that we have 5 black sheep. Since there are only two colors of sheep, the rest of our sheep must be white, although we can count them as well, just to be sure. Now we have three numbers, all from counting. We have 12 sheep, 5 of which are black and 7 of which are white.

Ratios express the relationship between two counts. They are exact measures of just how much bigger or smaller one number is than another.

Ratios are expressed as numbers in three principal ways, as proportions, as percentages, and as odds. Suppose we want to express the number of black sheep, n, in terms of its relation to the total number of sheep in our flock, N. The simplest way to do this is with an odds. We subtract the number of black sheep from the total number of sheep to obtain the number of sheep that are not black. The odds are expressed with the count of the items of interest, n, followed by a colon (:), followed by the remainder, $N–n$, shown in Equation 3-1.

$$5 : 7 \qquad\qquad\qquad (3\text{-}1)$$

In ordinary language, we would say that the odds that any one of our sheep we come across will be black are five-to-seven. If we express it in terms of chances, rather than in terms of odds, we say that the chances are five-in-twelve or five-out-of-twelve.

PROPORTIONS AND PERCENTAGES

Note that when we use the chances terminology (five-in-twelve instead of five-to-seven), we do not use subtraction. We state the number of black sheep directly in terms of the total number of sheep, which was our original goal. These two numbers are the basis for the other ways of expressing ratios, as proportions or as percentages. Both of these measures are calculated using division. To calculate a proportion, we take the count of the objects of interest and divide by the total number of objects, as shown in Equation 3-2.

$$p = n/N = 5/12 = .417 \qquad\qquad (3\text{-}2)$$

A couple of things to note about proportions. First, a proportion is a single number calculated from two other statistics. Second, when calculated in this way, with the first number being the count of just those subjects of interest and the second number being the total number of subjects, a proportion can only be between zero and one. If we had no black sheep, the proportion would be zero. If all our sheep were black, that is, we had 12 black sheep, then the proportion of black sheep would be one.

A percentage is just the proportion multiplied by one hundred. Percentages are sometimes used because they can be expressed as whole numbers, rather than as fractions or decimals. Also, when other sorts of ratios are taken that can be greater than one, percentages are more commonly used than proportions.

HANDY HINTS

Ratios Greater Than One

When can a ratio be greater than one? Only when the subjects of interest are not truly part of the total. This is common in the comparison of two counts taken at different times. For instance, if we breed our sheep this year and next year we have fifteen sheep instead of twelve, we might want to express the increase in our flock by comparing the two numbers as a percentage: $15/12 \times 100 = 125\%$. Next year, we would say that our flock was 125% of the size it was this year, or we could say we had a 25% increase in the size of the flock. Note: This is a ratio, but not a proportion. A proportion is a ratio of a part to the whole, and is therefore always between zero and one.

The most important fact about proportions is that probabilities, the numerical measures we use to express the likelihoods, are based on the mathematics of proportions. Like proportions, probabilities range from zero to one and the higher the probability, the more likely the event. Also, key to the theory of probability is the distinction between the ratio between the subjects of interest and the remainder of all of the rest, as is calculated via subtraction in the odds.

Note that, mathematically, if p is the proportion of subjects of interest to the total, then $1-p$ is the proportion of subjects not of interest. This is because the total population is comprised of exactly the subjects of interest, plus the subjects not of interest, as illustrated in Equation 3-3.

$$(N - n)/N = (12 - 5)/12 = 7/12 = .583 = (1 - .417) = (1 - p) \qquad (3\text{-}3)$$

The proportion of subjects not of interest, called the *complement* of the proportion of subjects of interest, is extremely important to the theory of probability.

Three Types of Probability

Traditionally, there are said to be three concepts of probability:

- *Classical probability*, which relies on the notion of equally likely events.
- *Frequentist probability*, which relies on the notion of replication.
- *Subjective probability*, which relies on the notion of rational choice.

Happily, as it turns out, all three notions are exactly the same, mathematically. This means that the distinction is primarily (and perhaps entirely) philosophical. Here, we will use the different types to show how probability relates to the kinds of events we need to know about for business decisions.

COUNTING POSSIBLE OUTCOMES: THE RULE OF INSUFFICIENT REASON FOR CLASSICAL PROBABILITY

The theory of probability was first developed to handle problems in gambling. The great mathematician, Pascal, was working to help out a gambler who wanted to know how to bet on games of chance, especially dice. In games, such as dice or cards, chances are easier to calculate, because everything can be counted. This allowed Pascal to work out the first theory of probability in terms of odds. This version of probability theory is called *classical probability*.

The rule of insufficient reason

The theory of probability is basically the application of the mathematics of ratios and proportions to the issues of chance and likelihoods. In one brilliant move, Pascal was able to bridge these two very different fields and create the theory of probability. Let's see how he did it.

Suppose we have a standard deck of cards, face down on a table, and we draw one card from the deck. What is the likelihood that we will draw the King of Hearts? Intuitively, we would say that the likelihood is low. After all, there are 52 cards in the deck and only one of them is the King of Hearts. What is the likelihood that we will draw the Eight of Clubs? Also low, and for the very same reason. Pascal then asked a critical question: Which is more likely, that we will draw the King of Hearts or that we will draw the Eight of Clubs?

Again, intuitively, since our reasons for assessing the likelihood of each is the same, there does not appear to be any reason to assume that either draw is more or less likely than the other. Pascal then proposed a new rule: Whenever we have no reason to think that one possibility is more or less likely than another, assume that the two likelihoods are exactly the same. This new rule is called *The Rule of Insufficient Reason*. (You can't beat the Renaissance thinkers for nifty names!) This one rule makes it possible to apply all of the mathematics of ratios and proportions to the problems of chance in gaming and, eventually, to all other likelihoods as well.

Measuring probabilities with proportions

We will get to the mathematical rules of probability a little bit later. For right now, it's enough to know a few important facts. Very little is needed to make the mathematics of probability work. In fact, only three basic rules are needed. See below.

TIPS ON TERMS

The Basic Rules of Probability
Scalability. The measures of probability must all be between zero and one.
Complements. The probability of something not happening must be equal to one minus the probability of that same thing happening.
Addition. For any group of events, the probability of the whole group must be equal to the sum of the probabilities of each individual event.

Collectively, these three rules are known as Kolmogorov's axioms, after the mathematician who discovered them almost 300 years after Pascal. Notice how well these rules fit in with a situation where we can count up all the events, as in the games of cards, or dice, or in counting sheep: proportions of subjects that have a particular property (like the color of the sheep or suits in a deck of cards) are all between zero and one. We have also seen how, in the case of sheep, the proportion of sheep that are not black (known as the complement) is one minus the proportion of black sheep. It looks like proportions may make good measures of probability.

This leaves the rule of addition. All that is left is to show that the sum of the proportions of different types of sheep (or cards or dice) is equal to the proportion of all those types taken together. If that is true, then proportions (which we already know how to calculate) will work just fine as our numerical measure of likelihood, which we call *probability*.

Let's expand our example of shepherding a bit. Suppose we have three breeds of sheep, heavy wool merinos, fine wool merinos, and mutton merinos. There are four heavy wool sheep, two fine wools, and six muttons. The proportion of heavy wools is 4/12. According to the rule of complements, the proportion of sheep that are *not* heavy wools should be $(1 - 4/12) = 8/12$. We don't need the rules of probability to count the sheep that are not heavy wools. There are eight, the two fine wools and the six muttons. Because the counts all add up—$(2 + 6 = 8)$—and the proportions are just the counts divided by 12 (the total number of sheep in the flock), the proportions add as well $(2/12 + 6/12 = 8/12)$. As we can see, so long as we can count all of the individual subjects, the rule of addition applies, too. And, when we divide by twelve, all of our figures can be expressed so that the measures of probability are between zero and one. As a result, we have met the basic mathematical requirements of probability, and we can apply the laws of probability to our counting of sheep (unless it puts us to sleep).

CRITICAL CAUTION

The probability of an event, $Pr(x)$, is not a statistic. It is not a measure of a general property; it is a measure of a specific attribute of a single event. The *proportion*, p, is a statistic. When calculated from a sample, the proportion provides an estimate of the probability of a specific event, using information from the entire sample. In statistical theory, the proportion of the entire population is a theoretical model of the probability (at least according to some theories of probability).

Probabilities in the real world

The notion of equally likely probabilities is, like most elegant mathematical ideas, never true in the real world. It takes enormous amounts of technology to manufacture dice so that they are nearly equally likely to land on each of their six sides. Casino dice come with a guarantee (a statistical guarantee!) that they will come pretty close to this ideal. Casino dice cost a lot more than the dice we buy at a convenient store for just this reason. Playing cards have been around for centuries, but the current playing card technology is only about 50 years old. In no case are dice or cards or other human manufactured technologies absolutely perfect, so the assumption of equally likely outcomes is, at best, only an approximation.

In the case of gaming technologies, there is an explicit effort to create equally likely outcomes, in order to satisfy the assumption based on the rule of insufficient reason. In the rest of the world, even this assistance is lacking. Consider even our simplified flock of sheep. It is unclear even what it would mean to have an exactly equal likelihood of selecting one sheep in our flock over another. If we are pointing out sheep, smaller sheep might be harder to spot. If we are actually gathering them up, friskier sheep might be harder to catch. Even if sheep breeders are breeding sheep for uniformity, they are not doing so to help our statistics, and even if they were, there will always be more variability among sheep than among dice.

The rule of insufficient reason does not mean that we have good reason to believe that all of the basic outcomes (like one side of a die showing up, or one particular sheep being picked) are equally likely to occur. It merely says that when we don't have any reason to think that any two basic outcomes are *not* equally likely to occur, we can base our measure of probability on counting basic outcomes. In classical probability, these basic outcomes are called *simple events*.

Mutually exclusive events

Finally, there is an important concept that applies to all three types of probability, but is best understood in the case of classical probability. Note that we have been considering different values (black and white, or heavy wool, fine wool, and mutton) for only a single variable (color or breed) at a time. This was a trick to ensure that all of the events were what is called *mutually exclusive*. Two events (and the probabilities of those events) are mutually exclusive if the fact that one happens means that the other cannot possibly have happened. If the color of the sheep we pick is black, its color cannot be white. If the sheep is a mutton merino, it cannot be a heavy wool merino. This is always true for different values of a single variable.

Things get a bit more complex when we consider more than one variable at a time. If the sheep we pick is black, it might or might not be a fine wool merino. We can't really know unless we know the relationship between the colors and the breeds for our entire flock. If one or both of our two fine wool merinos is black, then the event of picking a black sheep is not mutually exclusive of the event of picking a fine wool merino. However, if it happens that both of our fine wool merinos are white, then picking a black sheep means we definitely did not pick a fine wool merino, and vice versa. The two events are mutually exclusive despite being defined by values on two different variables.

REPLICATION AND THE FREQUENCY APPROACH

What do we do when we have good reason to suspect that our most basic outcomes, the simple events, are *not* equally likely to occur? If our business is farming, we may want to know whether or not it will rain. Rainy days and sunny days may be our most basic events. We certainly cannot assume that it is as likely to rain on any given day as it is to be sunny. Climate, season, and a host of other factors get involved. We have very good reason to suspect that, for any given day, in any particular place, that the likelihood of rain is not equal to the likelihood of sunshine. In similar fashion, the likelihood of showing a profit is not the same as the likelihood of sustaining a loss. The likelihood of a job candidate having a degree is not likely to be the same as the likelihood that he will not. For some jobs, almost all prospective candidates will have degrees; for other jobs, almost none.

In cases such as these, we need a new rule for assigning values for our probabilities. This time the rule depends on Hume's assumption (discussed in Chapter 1 "Statistics for Business") that the future will be like the past, which is key to the philosophy of science, which provides the model for the second type of probability, based on the theory of *relative frequency*.

In science, the assumption that the future will be like the past leads us to assume that, under the same circumstances, if we do things exactly the same way, that the results (called the *outcome*) will come out the same. The basic idea behind a scientific observation or experiment is that things are done in such a very carefully specified and documented way that the next person who comes along can read what we have done and do things so similarly that she or he will get the same results that we did. When this is true, we say that the observation or experiment is *replicable*. Replicability is the heart of Western science.

Frequentist theoreticians have an imaginary model of the scientific experiment called the *simple experiment*. They define simple experiments in terms of gambling devices and the like, where the rule of insufficient reason applies and we know how to calculate the probabilities. Then they show that, in the ideal, simple experiments, repeated many times, will produce the same numbers as classical probability. The big advantage to frequentist probability is that, mathematically, simple experiments work even when the underlying simple events are *not* equally likely.

The first simple experiment that is usually given as an example is a single flip of a coin. Then the frequentist moves on to dice. (Trust us. Heads still turn up 50–50 and each side of the die shows up 1/6th of the time. Everything works.) We will skip all this and construct a simple experiment with our flock of sheep. Suppose we put all of our flock into an enclosed pen. We find someone who is handy with a lasso, blindfold her, and sit her up on the fence. Our lassoist then tosses her lasso into the pen and pulls in one sheep at a time. (Simple experiments are theoretical, and don't usually make much sense.) The lassoing is our model of *sampling*, which we learned about in Chapter 2 "What Is Statistics?" Importantly, after the sheep is lassoed and we take a look at it, we then return it to the flock. (Like we said, these experiments don't make much sense.) This is called *sampling with replacement*.

TIPS ON TERMS

Sampling with replacement. In the context of an imaginary simple experiment, an act that determines a single set of one value for each variable in such a way that the likelihood of the different values does not change due to the act of sampling itself. Examples are: the flip of a coin; the roll of a pair of dice; the drawing of a card from a deck of cards, after which the card is placed back in the deck.

Note that things like flipping a coin or rolling dice, which we might not ordinarily call "sampling" count as sampling in statistics. When we flip a coin, we are said to be sampling from the space of possible outcomes, which are the events, heads and tails. This is sampling from a set of abstract events, rather than from a set of physical objects. What makes it sampling with replacement is that, once you flip a coin, the side that lands up doesn't get used up for the next toss. In terms of the odds, nothing changes from one flip of the coin, or one roll of the dice, to the next. With cards, in order to keep the odds the same, we have to replace the card drawn into the deck, hence the expression, *with replacement*.

Sampling without replacement. In the context of an imaginary simple experiment, an act that determines a single value for each variable in such a way that the likelihood of the different values changes due to the act of sampling itself. Examples are: the drawing of a card from a deck of cards, after which the card is set aside before the next draw; choosing a name from a list and then checking off the name.

The vast majority of statistical techniques, and all that we will cover here in *Business Statistics Demystified* assume that sampling is done *with* replacement. Mathematically, sampling without replacement is very complicated because, after each subject unit is removed from the population, the size of the population changes. As a result, all of the proportions change as well. However, sampling with replacement does not make sense for many business applications.

Consider the example of surveying our customers: we have a list of customers and are calling them in random order. In order to sample with replacement, we would have to keep a customer's number on the list even after we'd interviewed them once. But if we do that, we might pick the exact same phone number again and have to call that same customer! ("Hi, Mr. Lee! It's me, again. Sorry to bother you, but I need to ask you all those same questions again.")

Of course, in these sorts of cases, sampling with replacement is never really done, but the statistics that are used assume that statistics with replacement is always done. The trick is that, mathematically, if the population is infinitely large, sampling without replacement works identically to sampling with replacement. If our population is finite, but very large compared to the total size of our sample, we can pretend that it is infinite, and that all sampling is sampling with replacement, without generating too much error.

What is the probability that we will lasso a black sheep? According to classical probability theory, it is 5/12. Let's have our lassoist lasso sheep 120 times, releasing the sheep afterwards each time. We will probably not find that we have lassoed exactly 50 sheep (equal to 5/12 times 120), but we will be pretty close. In short, we can estimate the true probability by repeating our simple experiment, *counting* the different types of outcomes (black sheep or white sheep, in this case), and calculating the proportion of each type of outcome. An advantage of frequentist probability is that it uses proportions, just like classical probability. The difference is that, where classical probability involves counting the different possible types of simple events and assuming that each is equally likely, frequentist probability involves repeating a simple experiment and counting the different outcomes.

HANDY HINTS

Later on, after we have learned some additional tools, we will see that the frequentists have a better way of performing simple experiments in order to estimate the true probability. Without giving too much away, let's just say that it turns out that it is better to do ten experiments, lassoing twelve times for each experiment, than doing one experiment lassoing 120 times.

Why is this difference important? The reason is that the outcomes of simple experiments don't have to be equally likely. If our simple experiment is to flip a coin or roll a die, the outcomes are heads or tails, or the number on the top face of the die, and the outcomes can safely be assumed to be equally likely. But what about our simple experiment lassoing sheep? If we think of the outcome as being which of the 12 individual sheep gets lassoed, then each outcome is equally likely. But, suppose we aren't on familiar terms with all of our sheep, and don't know them all individually? We can think of the outcomes as lassoing any black sheep and lassoing any white sheep. Unless we count all the sheep in our flock and apply classical probability, we don't know what the relative likelihoods of lassoing a black sheep or a white sheep are, and we certainly cannot assume they are equal. White sheep are vastly more common than black sheep, and this is a very good reason to assume the likelihoods of picking each type are *not* equal.

There are two sorts of cases where it is good to have the frequentist approach, one where classical probability can be very hard to apply, and one where it is impossible to apply.

First, suppose we had a huge flock of sheep. We aren't even sure just how many sheep we have. We want to know the probability that we will pick a black sheep. If we define the outcome of our experiment as "black sheep" and "white sheep," we can estimate the probability of picking a black sheep without having to count our entire flock, or even being able to tell one sheep from another, except for their color. This illustrates both the convenience of the frequentist approach and the power of the sampling upon which it depends.

Second, so long as we can construct a simple experiment to sample some attribute of our subjects, we can estimate the probabilities. This is very useful in cases like the weather, profits and losses, and level of education (discussed above), where we have no way of counting anything except the results of sampling. Often, we do not even have to be able to conduct the simple experiment. (No blindfolds required!) We can just collect the data for our statistical study according to best practices, and treat the numbers as if they were the outcomes of simple experiments. This illustrates how probability based on relative frequency can be very useful in real world statistical studies.

COMMON SENSE AND SUBJECTIVE LIKELIHOODS

Classical probability and frequentist probability are typically classified together as types of *objective* probability. Here, "objective" means that there is a set of rules for calculating the precise numbers that does not depend on who actually does the experimenting or the counting or the calculations. (Note that this is a very different meaning for the word "objective" than is used in other contexts.) If it matters who does the work that produces the numbers, then the probabilities are called *subjective*.

There is also a mathematical theory of subjective probability, which has the same advantages over frequentist probability that frequentist probability has over classical probability. Subjective probability can be applied in cases where not only can we not count things, but where we cannot even expect things to be repeatable.

A good example might be a civil law suit. The details of every lawsuit and the peculiarities of every jury might be so dramatic as to prevent any sensible notion of repeatability. If we are in the business of manufacturing widgets and several other widget manufacturers have been sued for sex discrimination, or the issues in the lawsuit for widget manufacture are similar to those that have been raised in automobile manufacture, then frequentist probability might apply. But if we are the first widget manufacturer to be sued for sex discrimination and the widget business is importantly different than other businesses with regard to the legal issues for sex discrimination, then frequentist probability may not be useful. The only way we have of estimating the probabilities would be to call in an expert who knows about both widgets and sex discrimination lawsuits and have them make an educated guess as to our chances of winning the case. And this is just what subjective probability assumes.

TIPS ON TERMS

Subjective probability is also called Bayesian probability, because estimating the true values requires an equation called *Bayes' rule* or, more extravagantly, *Bayes' Law*. The name Bayesian probability is a bit misleading, because Bayes' Law can be applied to any sort of probability.

What subjective probability requires in place of replicability or the rule of insufficient reason, is a gambling game and players who are too sensible to get cheated. The game has the same purpose in subjective probability that the simple experiment has in frequentist probability. We imagine a game in which players bet on the outcome of some event. The game can be simple or complex. Remarkably enough, the rules of the gambling game do not matter, so long as it is fair and the players all understand the rules (and real money or something else of value is at stake). The event does not have to be repeatable, so long as the gamblers can, in principle, play the game over and over again, gambling on other non-repeatable events.

Being sensible is called being *rational*, and it is defined mathematically in terms of something called Decision Theory. It turns out that, if a player is rational in this formal sense, then his/her purely intuitive, subjective estimates of the probabilities (expressed as numbers between one and zero, of course) will not only satisfy Kolmogorov's axioms, but will also approximate the frequentist probabilities for repeatable events! Even more bizarre, if the player's initial estimates are off (presumably due to lack of knowledge of the area) and the player is rational about learning about the world during the playing of the game, his/her estimates will get better over time, again approaching the true probabilities.

CRITICAL CAUTION

It would be a big mistake to think that just because subjective probability can be applied more widely than frequentist probability and that frequentist probability can be applied more widely than classical probability, that subjective probability is somehow better than frequentist probability or that frequentist probability is better than classical probability. Each of the three types of probability requires different assumptions and there are always cases where some of these assumptions do not apply. We have seen where the rule of insufficient reason does not apply and we cannot use classical probability. When we cannot, even in principle, specify how something could be repeated, we cannot use frequentist probability. Subjective probability actually requires *seven* separate assumptions (called the Savage Axioms, after the mathematician L. J. Savage, who invented them), all of which are complex and some of which are controversial. There are cases where none of the assumptions hold and any notion of probability is suspect.

Using Probability for Statistics

We have seen what probability is. Now we will see some of how probability gets involved with statistics. We will learn about several key statistical concepts that require probability for a full understanding. We will see how probability is involved in how statistics deals with issues of causality, variability, and estimation.

STATISTICAL INDEPENDENCE: CONDITIONAL AND UNCONDITIONAL LIKELIHOODS

The very important concept of *statistical independence* is based on a relation between probability measures called *conditionality*. These concepts are important in using statistics to determine the causes of various facts in the world.

Finding causes

Understanding causality is a profound and difficult problem in philosophy. At best, statistics has a limited ability to detect possible cause–effect relations. However, statistics is one of the few techniques that can provide reliable information about causal relations at all. In short, when it comes to figuring out the cause of something, it is a limited tool, but, in many situations, it is the best tool we have.

It should go without saying that the information needed to make a business decision may very often not be information about a cause–effect relation. After all, it is lot more important to know that 90% of women between age 19 and 34 want to buy your new product than it is to know precisely what caused that fact. It should go without saying, but, unfortunately, it does not. Much of statistics comes from work in the sciences, and, in particular, the social sciences, where understanding cause–effect relations is taken to be of utmost importance. Because of this, statistics texts often spend a great deal of time focused on techniques for establishing cause–effect relations without even explaining why cause–effect relations are important, much less taking the time to consider when, in business, other sorts of statistics providing other sorts of information, may be more important.

FUN FACTS

The basic strategy for detecting the true cause of something observed in the world is called *Mill's method*, named after the philosopher, John Stuart Mill. Mill's method is actually five methods. The Method of Agreement means checking to see that the proposed cause be present when the effect is observed. The Method of Difference means checking to see that when the proposed cause is absent, the effect is absent. The Joint Method of Agreement and Difference involves checking groups of potential causes, systematically adding and removing potential causes, until one is found that is present and absent together with the effect. The Method of Concomitant Variation means checking to see that a proposed cause of more or less intensity results in an effect of more or less intensity. The Method of Residues means eliminating other possible causes by noting the presence of their separate known effects together with the effect of interest. Statistics takes a similar approach, with similar strengths and weaknesses.

From a statistical perspective, we would expect an effect to be more or less likely when the cause is present or absent. In order to look for causes, we will need a mathematical definition of the probability of one event when some other event has or has not happened.

Conditional likelihoods

Up until now, we have only considered the probabilities of individual events. These are called unconditional probabilities. The unconditional probability of event, A, is symbolized as $Pr(A)$. If we want to work with causal relations, we need to be able to talk about the relationship between two events, the cause, B, and the effect, A. For this, we use *conditional probabilities*. Let's look at an example:

The business cards for all nine Justices of the U.S. Supreme Court (as of 2003) have been placed face down on our desk. The probability of picking the card of a female member of the court, $Pr(Female)$, is 2/9. But suppose that someone picks a card, looks at it without showing it to us, and tells us that it is the card of a Republican member of the court? Knowing that the card is a Republican's, what is the probability that it is a woman's? In probability, we use the term *given* to express this relationship of conditionality. We ask: What is the probability of picking a woman's card, given that it is a Republican's? This is symbolized by $Pr(Female \mid Republican)$. Because only one of the female members of the Court is a Republican, and seven members of the Court are Republican, the probability, $Pr(Female \mid Republican) = 1/7$.

The mathematical rule for calculating the conditional probability for any two events, A and B is:

$$Pr(A|B) = Pr(A \& B)/Pr(B) \tag{3-4}$$

In order to see why this equation works, we can check our example. The probability of picking a card, from out of the original stack, of a Justice who is *both* female and Republican, $Pr(Female \& Republican)$, is 1/9. The probability of drawing a Republican's card, $Pr(Republican)$, is 7/9. And $1/9 \div 7/9 = 1/9 \times 9/7 = 1/7$.

CRITICAL CAUTION

Note that the probability of A given B, $Pr(A|B)$, is not the same as the probability of A and B, $Pr(A \& B)$. In terms of countable subjects, the probability of A given B only considers those subjects that are B. It is as if we are using only a part of the original whole population as our population, the part for whom B is true. The probability of A and B refers to a selection made from the entire population, not just the part of the population for whom B is true.

SURVIVAL STRATEGIES

One trick for remembering the equation for conditional probability is that the conditional probability is based on selecting from the smaller group where B has also happened. This means that the denominator must be changed from the total for the entire population to the subtotal for just the B's. Dividing by the proportion of the B's replaces the total with the subtotal. (The proportion of B's is just the subtotal of B's divided by the total, which is why it always works.)

The relationship of conditionality works for all sorts of events, not just those that are causally related. In fact, the two events we have been considering, drawing a woman's card and drawing a Republican's card are not even necessarily separate events, at least not in the ordinary sense. When a Republican woman's card is picked, that single action (in ordinary terms) is both the picking of a Republican's card and the picking of a woman's card at the same time. It all depends on how you describe it.

This is an important point for understanding probability. A single event, described in two different ways, will often be treated by probability theorists as two different "events," using their terminology. So long as the equations give the right answer, the mathematician and the theoretical statistician will be unconcerned. The trick to understanding this is that, when the equation works for any *A* and *B*, it will work for two events in the ordinary sense, just like it works for one.

Of course, if we are going to talk about causality (don't worry, we will get there soon), we have to talk about two events in the ordinary sense, because the cause has to happen before the effect. When we draw one card from the stack of business cards, the fact that we drew a Republican's card can't be the *cause* of the fact that that same card is also a woman's card. So we need an example where an earlier event can affect the probabilities of a later event. Recalling our definition of *sampling with replacement*, we know that, by definition, an earlier sample cannot affect the odds for a later one. So that sort of example won't do. And sampling without replacement is much too complicated. Here's a simpler example:

The old rule for eating oysters is to eat them only in months spelled with an 'R'. (This is due to the warm weather, so it's not true in the Southern Hemisphere.) Let's borrow our lassoist shepherdess for a moment, since she is already blindfolded, and have her throw a dart at a calendar in order to pick a month so that every month has the same chance of getting picked. The probability that she hits a month where oysters are safe, *Pr(safe)*, is 8/12. The probability that she hits a month where they are unsafe, *Pr(unsafe)*, is 4/12. The probability that she hits a month where they are safe *and* they were unsafe the previous month, *Pr(safe & unsafe)*, is 1/12. (The only month where this is true is September.) The probability that she hits a safe month *given* that the previous month was unsafe, *Pr(safe|unsafe)*, is 1/4. This is because there are five unsafe months, any one of which can be the previous month to the month picked, but only one of them, August, is followed by a safe month. *Pr(safe | unsafe) = Pr(safe & unsafe) / Pr(unsafe) = 1/12 ÷ 4/12 = 1/12 × 12/4 = 1/4*. So the rule for conditional probabilities also works for events where one event happens before the second.

What is a random variable?

We have been careful not to use the word "random" too much up to this point, because, both in statistics and in ordinary English, the word can mean more than one thing. Instead of having our shepherdess blindfolded and lassoing or throwing darts, we could just have said, pick a sheep or a month "at random," but that phrase is ambiguous. Sometimes "at random" just means "unpredictably." What was necessary for our examples is that each subject have an equal chance of being selected (our definition of a random sample) and that sampling be done with replacement, so that taking one sample doesn't change the odds on any later sample. So, instead of just saying "at random," we were very precise (and silly) about how things got picked. Being precise about specifying how samples are (or should be) taken is extremely important throughout statistics, as we will see at the end of the next section, on statistical independence.

In statistics, the word "random" also has two uses. In the phrase "random" sample, it means that everything has the same chance of being selected. But there is also a concept in theoretical statistics called a "random variable" and, here, the word random means something quite different.

A random variable is the way that statistical theorists use to talk about the ordinary variables we have seen that measure our subjects in mathematical language. Random variables can be defined either in terms of classical or frequentist probability. Technically, a random variable gives a number for each simple event (in classical probability) or for each outcome of a simple experiment (in frequentist probability). For example, we could assign the number 1 to each of our fine wool sheep, 2 to each heavy wool sheep, and 3 to each mutton. This provides a convenient mathematical way to talk both about events and data. Since we can calculate the probabilities for each event (or outcome), we can link each probability to one of the three numerical codes. We could call this random variable, *breed*.

In the case of a numerical measurement, such as Judy's height, the purpose of a random variable is clearer. Let's expand this example to include some of Judy's friends. The theoretical model of the data variable, *height*, is the random variable also named *height*. The random variable called height assigns a number to each person in the population of Judy and her friends, that happens to be the same as the number we get when we measure that person's height. In terms of data measurement, we pick Judy (or Tom, or Ng) and measure their height and get a number, 62 inches (or 71, or 63). In terms of statistical theory, we write: $height(Judy) = 62$. (Note that inches, the units of measurement, are not part of the value of the random variable.) And we can do this the same way for every measure we take of our subjects. For instance, $sex(Judy) = 1$ (for female), $sex(Tom) = 2$ (for male), $age(Hassan) = 20$, and $yearsOfEducation(Ng) = 13$ (Ng is a college sophomore).

More generally, the concept of a random variable allows us to deal with combinations of simple events (called complex events) and describe their probabilities in a mathematically convenient way. We leave off the name of the subject to indicate that we are talking about a sample from the entire population and write: *sex* = 1 to indicate the complex event of selecting any one of Judy and her friends who is female or *age* < 21 to indicate the complex event of selecting any one of Judy and her friends who cannot drink legally. We can even do something odd like writing: *breed* < 3 to indicate the complex event of selecting any one of our wool merinos. (It is probably safer to indicate this complex event by writing *breed* = (1 or 2), because the variable *breed* is nominal, not ordinal.)

Now that we can describe these events conveniently (and with less possibility of ambiguity), we can improve our notation for probability: $Pr(breed=1) = 2/12$ and $Pr(height<64) = 7/24$.

HANDY HINTS

The use of random variable notation for specifying probabilities shows us another difference between probability theory and statistical theory. In the notation, $Pr(variable=value)$, the formula inside the parentheses names an event. In probability theory, events are defined in practical, real-world terms and are the basis for all of the other definitions. In statistical theory, events are defined in terms of specific variables having specific values. This fits in exactly with Thomsett's (1990) point about the difference between probability and statistics. In probability, we need to understand the process (like rolling dice or lassoing sheep) to define events. In statistics, events are defined strictly in terms of the results.

Statistical dependence and independence

At last, we now have enough background to talk about how statisticians detect possible causal relations. If the likelihood of an event is not different due to the presence or absence of an earlier event, then that earlier event cannot be a cause of the later event. At a very minimum, the presence or absence of a cause should make the effect somewhat more likely or unlikely. This relationship is defined in terms of *statistical dependence*.

Statistical dependence or independence is a property that can apply either to events or to random variables, but it means something different in each case. These differences relate to the way we might ask questions about causes. Suppose we are interested in knowing whether the weather affects sales in our industry. This is a question about variables. The weather can have many values (sunny, rainy, hot, cold, etc.) and sales can have at least three (up, down, or steady). On the other hand, we might want the answer to a more specific question, such as whether sales go down on rainy days. This is a question about events. We are asking about the relationship between when *weather* = "*rainy*" and *salesTrend* < 0. Statistics can help us in both cases, but we have to be clear about our question.

For events, the definition of independence goes as follows: Two events, *A* and *B*, are independent if $Pr(A) = Pr(A|B)$. In other words, if the probability of *A* is the same whether or not *B* happened (and even whether or not we know if *B* happened), then *A* is *independent* of *B*.

The importance of statistical independence is that it is the minimum guarantee we can have that there is *no* causal relationship between two events. It may be hard to detect causes, but if the first event makes no difference to the second event, we can be as sure as we can ever be that that first event was not the cause of the second. Statistical independence is easier to define, but harder to find, than statistical dependence.

HANDY HINTS

In statistics, the focus is almost always on statistical independence, rather than on statistical dependence. The reason for this is that almost all statistical techniques require certain things (usually variables) to be statistically independent. These requirements, of course, take the form of *assumptions* for this or that technique, and are thus of the utmost importance in choosing which statistical technique to use. What tends to get left out of these discussions is *why* all of these statistical techniques require statistical independence. Remember that almost all of these techniques are intended to help identify cause–effect relations, which show up as statistical dependencies. In order to be able to spot causal relations, everything about the design of our study and the organization of our data must be free from statistical dependencies, so that any statistical dependencies that show up in the results are due to what was found in the study. We need to have statistical independence almost everywhere in order that the true underlying causal relationships show up as the only statistical dependencies in a field of independence.

In order to ensure the complete absence of dependencies between events required by statistical procedures, statisticians have defined statistical independence between random variables. Two random variables are statistically independent if every event defined in terms of any individual values for the first variable is independent of every event defined in terms of any individual values for the second variable. For instance, it might be that rainy weather doesn't decrease sales, but that cold weather does increase sales. The event, sales decrease, is independent of the event, a rainy day, but the variable, sales, is dependent on the variable, weather. In short, if there is any dependency anywhere between events described using the two random variables, then the random variables are not independent.

TIPS ON TERMS

The distinction we are using here, between the theoretical notion of a random variable and the practical concept of a measurement variable, is not usually defined in statistics textbooks, but it does reflect the way statistics is done in the real world. In *Business Statistics Demystified,* "random variable" is a theoretical notion related to populations and "variable" is a practical notion related to samples.

CRITICAL CAUTION

Statistical dependence is not the same thing as a random sample (defined as every individual unit having an equal chance of being sampled). It is entirely possible to have a non-random sample where everything is entirely statistically independent. In rolling a pair of dice, each roll is independent, but snake eyes is less likely than an eleven. (In classical probability, where we define each simple event uniquely, five on the first die and six on the second is a separate event than six on the first die and five on the second, this confusion doesn't usually happen. But in frequentist probability, it is entirely possible. Remember, in actual business practice, we can't always define simple events and must use frequentist probability.) We may very often want to have certain sampling events more likely than others, but we will almost never want to have one sampling event change its likelihood due to a previous one.

Errors in sampling

The most important reason that statistical independence is required by so many statistical procedures is that really horrible errors can happen when statistical dependencies creep into our methods. The examples of these errors are legion. One of the worst, and most common types of errors arises when we cannot use a truly random sample, because things are happening sequentially in time in an order that we cannot control. We can see an extreme case of this in our example of the safe and unsafe months for eating oysters (in the preceding section). Suppose that the safe and unsafe months were scattered about throughout the year. Then the likelihood of a safe month following an unsafe month would be far greater than it is. The reason that this probability is so low is that (a) safe and unsafe months cluster together and (b) unless we have a calendar, some darts, and a blindfold, we have no control over the order in which the months appear. The oyster example is an extreme one (by design), but this type of problem occurs all too often in real life in ways severe enough to cause real problems for statistics.

Let's return to our example, from our definition of systematic sampling (in Chapter 2 "What Is Statistics?") of taking every tenth product unit off of the assembly line for testing. If defects in manufacture tend to cluster, then taking every tenth unit may work, so long as the clusters are either much bigger than, or smaller than ten. If defects tend to cycle rhythmically, showing up regularly after every so many good units, then we had better be sure that the cycle isn't a multiple of ten. For instance, if our stamping machine stamps out molds for our products 100 at a time, and the 47th one is always bad, because that part of the stamp is broken, then testing every 10th unit will not catch the problem. One possible way of avoiding these types of problems would be to wait until the units are in bins and pick our test samples from the bins. This gives us much more control over the order in which things are sampled. Even here, we have to be careful, as units that come off the assembly line one after another may tend to cluster together in the bin. We should pick a test sample from the bin, stir things about a bit, and then pick another, and so forth.

If things are this difficult in a nicely structured, organized place like an assembly line, imagine how much more difficult they are out in the world, where our customers and vendors are. Let's look again at the example of the Supreme Court. Suppose we are big fans of the Court and want to catch sight of the justices. We go to the court, find ourselves a place to sit in The Great Hall, and wait for a glimpse of each justice. There are some good things about this strategy. The Great Hall is on the main floor, as are all the justices' chambers. We can expect them to go back and forth while consulting with one another. It is even sampling with replacement, because if one justice crosses the Hall leaving her office, she may also be the very next justice we see, on her way back. In fact, this may be a problem, because having just left one's office may increase (or decrease) the likelihood that that same justice will be seen again soon. Worse, Republicans may spend more time with Republicans than with Democrats. So, when they gather to go to lunch, seeing a Republican first may mean it is more likely that the next person we see will be a Republican as well.

Of course, these statistical dependencies (whether or not they are the results of clustering) are only a problem if they interfere with getting useful information for our decision. A few examples will illustrate how bad these problems can get. First, there is the classic goof (discussed by Huff & Geis, 1954) by the *Literary Digest* (a magazine in olden days), which predicted that Republican Alf Landon would defeat the then-President Franklin Delano Roosevelt in a landslide in the election of 1936. (FDR won in a landslide.)

Part of the problem was that this political opinion poll was taken by telephone. Back in 1936, at the height of the Great Depression, a lot of folks couldn't afford phones, and those people didn't much like the Republicans, whom they blamed for the Great Depression. If only people with telephones voted, Alf Landon might have become President.

Another big problem is the post-hoc hypothesis, where someone decides what causal relationship to look for after they check the data. The problem here is that, after the data are collected, the conditional probabilities change. Suppose that we are playing cards and we think that someone is cheating and that the deck is stacked. We predict that if a third party turns over the top card, it will be the Jack of Spades. Someone steps up and turns over the card and, lo and behold, it is the Jack of Spades. If the deck were not stacked, the odds on our making that prediction successfully would be 51 to 1 against. Either we just got very lucky, or else the deck was stacked. Now, suppose, instead of making a prediction, we wait until the card is turned over. It is a Jack of Spades, which is bad for us and we lose the hand. *Then*, we say, "It's the Jack of Spades. I knew it! This deck is stacked!" Should any of the other players believe us? The likelihood that the deck is stacked given that the card we predict appears is very different from the likelihood that the deck is stacked given that a card we don't like appears. (Using conditional probabilities, this can be shown mathematically as well, but the calculations are complex and beyond the scope of this book.)

A common form of post-hoc hypothesis is the multiple comparisons problem. The problem here is that, with a lot of data, even when it is properly selected, some patterns are bound to show up. Suppose we measure all our sales in order to see if the weather affects sales. We sell hundreds of products. Instead of checking to see if overall sales change with the weather, we check the sales of each and every product to see if the sales change with the weather. And, lo and behold, the sales of Part #36503 match the weather exactly! This is a case of asking too many questions on just one topic. Ask enough questions and, just by sheer luck, the answer to one of them will be "yes."

The thing that all of these examples have in common is that the sampling procedure affects the probabilities of the events of interest. Sampling from the assembly line in tens increased the probability of missing certain defects. Using the telephone for the political opinion poll increased the probability of sampling Republicans. Waiting until we saw the Jack drawn (dramatically) increased the probability that we would say that the card was a Jack. Sampling each product's sales separately increased the probability that we would find at least one pattern that matched the weather. This is yet another reason why, when we do statistics, we must be very careful about how we sample and we must document the sampling procedure very precisely.

WHAT IS A PROBABILITY DISTRIBUTION?

Central to statistical theory is the notion of a *probability distribution*. Every random variable has a probability distribution which specifies the likelihood of each value for that variable occurring. Understanding the mathematical theory behind distributions is not especially important for doing statistics, but important decisions as to which statistical techniques to use to answer certain types of questions require knowing whether or not our data variables are distributed according to one or another theoretical probability distributions. As we will see in Part Three, most standard statistical techniques require that our data be distributed in a certain way. If it is not, an entirely separate set of statistical techniques, called nonparametric techniques (see Chapter 14 "Nonparametric Statistics") must be used instead. Therefore, we need to understand the basics of what a probability distribution is.

Sampling distributions

Probability distributions are theoretical aspects of random variables related to populations, but there are also distributions that are not theoretical at all. These are sample distributions, and they can be of enormous practical use in all sorts of statistical analysis. A sampling distribution is just the way the different values for one single variable vary over our sample. We will return to our examples of shepherding and Judy and her friends in order to illustrate all this.

First, let's look at the distribution of the breeds of our merino sheep in Fig. 3-1. The distribution of a categorical variable is easily shown with a bar chart (see Chapter 7 "Graphs and Charts" for more on bar charts).

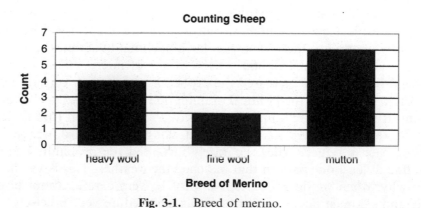

Fig. 3-1. Breed of merino.

Table 3-1 Heights of Judy and friends.

Name	Height (inches)
Judy	64
Angela	64
Betty	$66\frac{1}{2}$
Caroline	63
Donna	67
Francie	$69\frac{1}{2}$
Hannah	$63\frac{1}{2}$
Liora	68
Maria	$66\frac{1}{2}$
Ng	$63\frac{1}{2}$
Rita	64
Shizuko	$65\frac{1}{2}$
Tammy	61
Wai-Jung	$64\frac{1}{2}$

The order of the bars is arbitrary, because the variable, breed, is nominal and not ordinal. Because of this, there is very little more to say about the distribution. So let us move on to Judy and her friends.

First, we need to take a look at the distribution of heights of Judy and 13 of her female friends. Before we graph it, let's take a look at it in Table 3-1.

The best way to take a look at the distribution of this sort of data is to build a *stem-and-leaf* diagram. We start by taking all but the last digit of each number as the stem. Since our data are rounded to half inches, the fractional part (either .0 or .5) is the last digit. Here is our first try at a stem-and-leaf. First, we put all the numbers in a table, with the stems in the left column and the leaves in the right, as you can see in Table 3-2.

Table 3-2   Stems and leaves (unsorted).

Height Data, Split	
Stem	**Leaf**
64	.0
64	.0
66	.5
63	.0
67	.0
69	.5
63	.5
68	.0
66	.5
63	.5
64	.0
65	.5
61	.0
64	.0

Next, we sort the table, creating Table 3-3.

Then, we combine all of the rows with the same stem in Table 3-4, by stacking the leaves in the right-hand column and adding any missing stems. (Here the only missing stem is 62.)

As we can see, the stem and leaf lists all the data and, at the same time, creates a horizontal bar chart of the distribution. It is quick and easy and can be done by hand or by computer. Because the heights of these people are numerical values, these bars are ordered, so we can see how the number of people varies with their heights. There is only one person, Tammy, who is

Table 3-3   Stems and leaves, sorted.

Height Data, Sorted	
Stem	**Leaf**
61	.0
63	.0
63	.5
63	.5
64	.0
64	.0
64	.0
64	.0
65	.5
66	.5
66	.5
67	.0
68	.0
69	.5

under 5 foot 2. Most of the women are between 5 foot 3 and 5 foot 6. Three of the women are over 5 foot 6. A table could tell us this, but not so easily. Now we can see that most of the women are of middle-height, with a few shorter and a few taller. This is the famous bell curve we have heard so much about. It shows up clearly in the stem-and-leaf diagram.

If we turn the stem-and-leaf on its side and make it into a histogram with polygon, we can see the shape even more clearly in Figs. 3-2 and 3-3.

What makes a bell-shaped curve bell-shaped is that there are more of the middle values than of either high or low values. In nature, this is very

Table 3-4   Stem-and-leaf.

Stem	Leaf
61	0
62	
63	055
64	0005
65	5
66	55
67	0
68	0
69	5

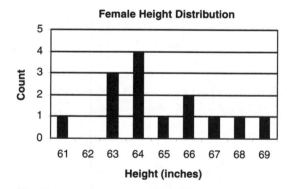

Fig. 3-2.   Female height distribution (bar graph).

common. Most people are not very short or very tall, nor very rich or very poor, etc. This is true for all sorts of units and for all sorts of variables. Most businesses are neither enormously profitable nor perpetually at the verge of bankruptcy. Most real estate properties are neither very large nor very small. Most stocks neither collapse in value nor soar. The bell curve describes this very general fact about the world.

As we can see from the figure showing the heights of Judy's friends, the curve of a distribution has all of the possible values of the variable along the

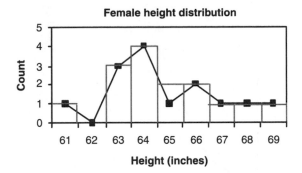

Fig. 3-3.   Female height distribution (histogram with frequency polygon).

horizontal axis (called the *x-axis*) and the count (called the *frequency*) of each value along the vertical axis (called the *y-axis*). Because the variable is numerical, the x-axis is just a number line, which you may remember from grammar school.

The numbers along the x-axis (called the *scale*) correspond to the values of the variable in question. The bump in the middle of the bell curve, which shows what values are most common, is called the *central tendency*. For the women in the graph of Judy's female friends, the central tendency is around 64 inches. If we graph some of Judy's male friends as well, we can see the first way that two different bell curves can differ. Men tend to be taller than women. In Fig. 3-4, we have placed the heights of Judy's male friends in gray next to those of her female friends. Looking at the graph, we see that the central tendency for the distribution of the heights of Judy's male friends is about 68 inches.

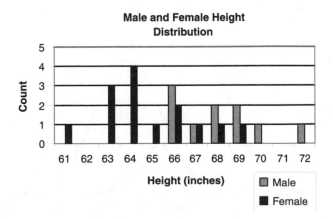

Fig. 3-4.   Male and female height distribution (bar graph).

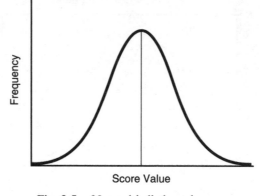

Fig. 3-5.   Normal bell-shaped curve.

We can also see that the distribution of the heights of Judy's friends does not look much like a bell. This is because there are only a few people being used to make up the numbers for the curve. Were we to graph the heights of 30 or 40 women, instead of just 14, we would find that the curve would be very much bell-shaped, like in Fig. 3-5.

In a case like that with the heights of males and females, we can have two bell-curves with exactly the same shape that have different central tendencies. In that case, the only difference will be the position of the curve along the x-axis, as in Fig. 3-6.

Even when there are lots of data, and the curves are smooth, there are many variations on the shape of the bell-curve found in real data. Another way that bell curves can vary is in how much the values stray from the central tendency. The weights of a large flock of mutton merinos might vary less than

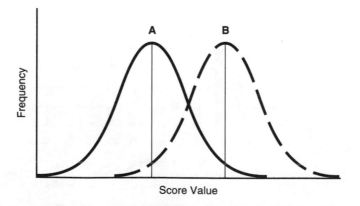

Fig. 3-6.   Two bell-curves with different means.

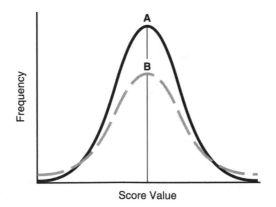

Fig. 3-7.   Two normal curves with different variances.

the weights of an equally large number of wild goats, because the mutton merino is bred to produce as much mutton as possible. Sheep tend to be pretty uniform in size. In Fig. 3-7, we can see two bell curves with the same central tendency, but with different amounts of *variability*.

As we saw in the case of the heights of Judy's friends, bigger values can be more common than smaller values, or vice versa. More of Judy's female friends were taller than 5 foot 4 than were shorter than 5 foot 4. If something has caused lower values or higher values to be more common, we will see curves like those in Fig. 3-8, called *skewed left* (or negatively skewed) and *skewed right* (or positively skewed, respectively).

The size of real estate lots, measured in acres, will be skewed left, because bigger lots are more expensive and get broken up into little ones, and because real estate lots can only get so small. The incomes of unionized workers may

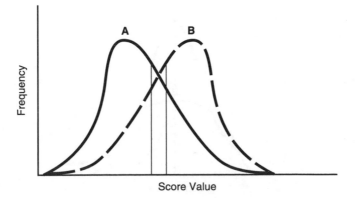

Fig. 3-8.   Two skewed distribution curves.

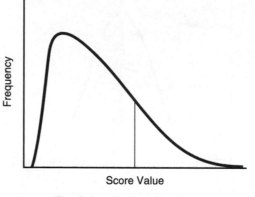

Fig. 3-9.   Truncated curve.

be skewed right, because the highest paying jobs in the firm are non-union. If the salaries of all employees fall into a bell-shaped curve, when we consider only the unionized employees, we are, in essence, chopping the bell-shape somewhere near the middle, which can create a severe skew. This type of curve in Fig. 3-9 is called a *truncated* curve.

Very often, we will find that there are slightly more very high and very low values than in a true bell curve. This causes the bump in the middle of the bell curve to be slightly flattened. The amount that the middle of the bell curve is sharpened or flattened is called the *kurtosis*, as seen in Fig. 3-10.

Theoretical distributions

As we can see, a distribution is just the way of showing how many times each value occurs for a single variable. By ordering the values numerically, we get

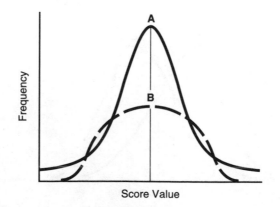

Fig. 3-10.   Two curves with different kurtotic indices.

a sense of how larger and smaller values are more or less common. The distributions above are sample distributions, because they show the *frequency* of each value that is actually found in our sample. If we were dealing with a population, particularly a population of unknown size or infinite in size, we wouldn't know exactly how many subjects would have each value for the variable. This is where the notion of a random variable comes in. For a population, instead of the exact count of each value, we will need to know the *probability* that each value will occur.

A theoretical distribution for a variable is a curve with all of the possible values of the variable along the x-axis and the relative probability of each value along the y-axis. The height of the curve of a theoretical distribution is re-scaled so that the total area under the curve is exactly equal to one. Together, these two features mean that the area under any part of the curve is exactly equal to the probability that the variable will have the values that appear within that part of the curve.

Theoretical distributions are built from mathematical equations that are designed to match the shapes of the real sample distributions of various kinds of data. The most common formula is the *normal* distribution, which has the true bell-curve shape. Figure 3-11 shows the standard normal curve, which has a central tendency of zero and whose variability is standardized to a value of one using a unit called *sigma* (symbolized by "σ").

The exact formula for the curve normal distribution is too complicated for us here, and we can manage just fine without it. It is important to understand that the normal curve is defined by its specific formula and that having an exact formula for the normal curve gives it specific features that are vital to understanding statistical inference, and therefore applies to all the

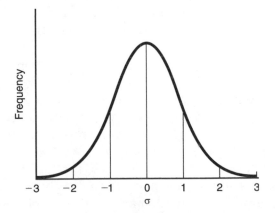

Fig. 3-11.   Normal distribution.

conclusions about consequences of business decisions drawn from statistics (the subject of Part Four of *Business Statistics Demystified*).

Because the normal curve has a specific shape, the proportion of the area under each part of the curve is always the same, no matter what the specific values of the population distribution are. In the standard normal curve, these areas are marked out in units of *sigma*. If we know any two points on the x-axis under the standard normal curve in terms of sigma, we know the probability that the variable will have a value between those two points.

By adjusting the standard normal curve to the central tendency and the variability of our data, we can use the shape of the normal curve to link values of a variable to the probability that those values will occur. We slide the curve to the right or left (as in Fig. 3-8) to match the central tendency of our data. Then we widen or narrow it (as in Fig. 3-9) to match the variability. Now we can link any value of the variable to the corresponding sigma value.

Suppose we handle financial transactions for the auto parts industry. We know that the price per sale is normally distributed. Small sales and big sales are costly to handle, so we want to offer a special rate to new customers for all transactions between $200 and $2000. In order to set the special rate, we need to know what proportion of sales will fall between these two values. If we know the central tendency and the variability of the price per sale, we can use the shape of the standard normal curve to calculate the probability of a sale being covered by our special rate. Even without looking at sales records, we can set our special rate.

This is the genius of statistical inference. Using distribution curves, we can associate real values measured from the real world with probabilities, which help us answer questions about the likelihoods of events, including future events, important to us.

In fact, the general technique shown above is even more powerful than that. So long as we know the precise mathematical form of the curve of the distribution of the variable in the population, that distribution does not even have to follow the normal curve. There are dozens and dozens of mathematical formulas for other sorts of distributions. And, for each mathematical formula, there is a way of calculating a probability and associating it with every point along the number line.

The trick is that we can never know for sure what the population distribution is. All we can ever really know is the sample distribution. There is a mathematical proof, called the *Law of Large Numbers*, which assures us that, so long as we take a big enough sample, the statistics for that sample will be close to the statistics for the population distribution. That means that we can look at the sample distribution and be reasonably confident that it is close to the population distribution. Then, if the sample distribution matches the shape of the normal distribution, or some other mathematically understood distribution, we can use the technique above to make inferences.

And the above approach is even more powerful than that. Even if the sample distribution is so oddly shaped that it doesn't match *any* distribution we know of, there is a less powerful form of statistical inference, called *nonparametric statistical tests* (discussed in Chapter 14 "Nonparametric Statistics") that works without the mathematical formula of the population distribution.

We can now see a little bit of how the logic of statistical inference relates to probability. The most important thing to know is that statistics works whether or not we understand statistical theory and probability theory. We need to understand the logic of statistical inference in order to do quality statistics. We only need to know statistical theory and probability theory if it is on the test.

The Laws of Probability

This section summarizes the basic terminology and rules that underlie probability theory and the laws of probability, which say what sorts of calculations can be used to determine complex probabilities from simple ones.

TIPS ON TERMS

Here are a few definitions of the basic terms from probability theory.

Simple event. One of a specific set of types of events. Simple events are defined to be *mutually exclusive*, which means that, on one specific occasion, only one simple event from the set can occur.

Complex event. A single event defined in terms of more than one simple event. Rolling an even number on a single die is a good example. This event occurs whenever any one of the three simple events (rolling a two, four, or six) occurs.

Complementary event. Also called the complement. A single event defined in terms of another event not happening. The complement of a simple event may be complex and vice versa. (The complement of rolling a six on a single die is the complex event of rolling any one of 1, 2, 3, 4, or 5.)

Joint event. An event that has two or more characteristics. For example, a red king is a joint event because the card is red *and* it is a king.

Simple experiment. A theoretical procedure that can be described in enough detail and that involves processes that are not too variable, so that when performed multiple times on different occasions, will tend to have the same proportion of different types of outcome on each occasion.

The outcome. One of a specific set of types of outcomes of a simple experiment.

Mutually exclusive events. Two events (and the probabilities of those events) are mutually exclusive if the fact that one happens means that the other cannot possibly have happened.

There are a few simple rules that define how probabilities are measured.

TIPS ON TERMS

The range. Probabilities are scaled from zero to one, in real numbers. This is mostly due to convention, but it is used universally because it makes the math easier.

The null event. An event that has no possibility of happening in the specific situation defined. For the null event, the probability is zero.

The certain event. An event that is certain to happen in the specific situation defined. For the certain event, the probability is one. The complex event corresponding to all simple events is a certain event.

There are a few simple rules that define how probabilities are combined.

TIPS ON TERMS

The addition rule. When more than one mutually exclusive event can happen, the probability of the complex event corresponding to all of those mutually exclusive events is equal to the sum of the probabilities of all of the mutually exclusive events.

The multiplication rule. When more than one independent event can happen, the probability of the joint event is the product of the probabilities of the two independent events.

Quiz

1. For understanding business statistics, the most important way that probability fits in with statistics is that it shows us...
   (a) The way that numbers are calculated from the population relate to the sample.
   (b) The way that the numbers calculated from the sample relate to the population.
   (c) Both (a) and (b)
   (d) Neither (a) nor (b)

2. The *Rule of Insufficient Reason*, which states that when we have no reason to think that one possibility is more or less likely than another we can assume that the two likelihoods are exactly the same, is related to which type of probability?
    (a) Classical probability
    (b) Frequentist probability
    (c) Subjective probability
    (d) We cannot calculate probability

3. Which of the following is the advantage to Frequentist probability and their use of the concept of the simple experiment over Classical probability?
    (a) Simple experiments work even when the underlying simple events are equally likely
    (b) Simple experiments work even when the underlying simple events are mutually exclusive
    (c) Simple experiments work even when the underlying simple events are not equally likely
    (d) There is no advantage of Frequentist probability over Classical probability

4. If we have a case where we cannot count things and we cannot expect things to be repeatable, which type of probability should we apply?
    (a) Classical probability
    (b) Frequentist probability
    (c) Subjective probability
    (d) We cannot calculate probability

5. Which of the following statements illustrate the importance of statistical independence?
    (a) It is the minimum guarantee we can have that there is no causal relationship between two events
    (b) It is the maximum guarantee we can have that there is no causal relationship between two events
    (c) It is the minimum guarantee we can have that there is a causal relationship between two events
    (d) It is the maximum guarantee we can have that there is a causal relationship between two events

6. Which of the following is *not* a problem of introducing statistical dependencies between variables?
    (a) They introduce sampling errors
    (b) They introduce the problem of post-hoc hypotheses
    (c) They cloud our ability to spot real causal relationships
    (d) All of the above are problems

7.   What is the value of organizing data in to a stem-and-leaf diagram?
     (a)   It allows us to see a rough estimate of causal relationships
     (b)   It allows us to see a rough estimate of the sampling distribution
     (c)   It allows us to see sampling error
     (d)   It allows us to see probability

8.   If two bell curves differ in their central tendency, the curves will differ in which of the following ways?
     (a)   The skew of the curves
     (b)   The truncation of the curves
     (c)   The position along the x-axis
     (d)   The kurtosis of the curves

9.   A sample distribution shows the _____ of each value that is actually found in the sample.
     (a)   Population value
     (b)   Organization
     (c)   Probability
     (d)   Frequency

10. In a theoretical distribution, the x-axis represents the _____ and the y-axis represents the _____.
     (a)   All possible values of the variable; the frequency of values in the data
     (b)   All possible values of the variable; relative probability of each value
     (c)   The range of values in the data; the frequency of values in the data
     (d)   The range of values in the data; the relative probability of each value

Exam for Part One

1. Which of the following does *not* lead to the underuse of statistics in business?
   (a) A failure to set up statistical systems in advance of decision making
   (b) A poor understanding of the limits of the available facts or the statistical techniques useful for converting those facts into information
   (c) Lack of statistical knowledge on the part of business persons
   (d) The time pressure to make business decisions

2. Which of the following does *not* lead to the overuse of statistics in business?
   (a) Mistaken assumptions about how complicated or difficult to use or costly statistics can be
   (b) Requirements made by bosses and standards organizations and legal authorities that fail to recognize limitations of statistics
   (c) A desire to justify a decision with the appearance of a statistical analysis
   (d) Failures by decision makers to determine the value of statistics as a part of their analysis

3. The key to knowing when to apply one statistical technique instead of another is to understand the _____ of the techniques.
    (a) Error
    (b) Statistical assumptions
    (c) Mathematics
    (d) History

4. Which of the following is *not* one of the three things we need to know, and can know, before statistics can be useful for a business decision?
    (a) We need to be able to characterize the current decision we face precisely
    (b) There needs to be a history of similar situations that we can rely upon for guidance
    (c) We need to know specific facts about the future outcome of our decision
    (d) There needs to be a history of similar decisions that we can rely upon for guidance

5. Which of the following is a question that can adequately be answered by descriptive statistics?
    (a) How many units failed quality checks today?
    (b) Did our lowest priced vendor give us enough of a price break on our key component last quarter to impact profits?
    (c) Have men or women purchased more of our product in the past year?
    (d) Do our employees rate their work experience more highly than do our competitors' employees?

6. Which of the following is a question that can adequately be answered by inferential statistics?
    (a) How many of our employees rate their work experience as very good or excellent?
    (b) How many women between 18 and 34 have purchased our product in the past year?
    (c) Which vendor gave us the best price on our key component last quarter?
    (d) Did enough units fail quality checks today to justify a maintenance call?

7. What are the advantages of conducting a statistical study over using a statistical technique on a one-time only basis?
    (a) It is cheaper to collect a lot of statistics at once that may help with a lot of decisions later on than to collect statistics one by one as they are needed
    (b) A single statistical study can collect and describe a large amount of information that can be used to help make an even larger number of decisions
    (c) Both (a) and (b) are advantages
    (d) Neither (a) nor (b) are advantages

8. Which of the following components of a statistical study is not necessary to present in a statistical report?
    (a) The calculations of the statistical techniques used in the statistical study
    (b) The circumstances of the statistical study
    (c) The assumptions of the statistical study
    (d) The results of the statistical study

9. Which of the following is not an advantage of understanding how to lie with statistics?
    (a) It is the best way to learn that sound judgment is vital to making statistics work for us
    (b) It allows us to create convincing advertising campaigns
    (c) It helps us to learn the strengths and limitations of statistical measures and techniques
    (d) It helps us to be cautious about incomplete measurements and undefined terms in statistical reports

10. What is the correct order of the first 3 steps in performing statistics?
    (a) Analysis, sampling, and measurement
    (b) Sampling, measurement, and analysis
    (c) Analysis, measurement, and sampling
    (d) Measurement, sampling, and analysis

11. Which of the following statements about measurement is not true?
    (a) Measurement is a formalized version of observation
    (b) Measurement is different from ordinary observation
    (c) Measurement provides a specific description of the world
    (d) Measurement provides a general description of the world

12. How is a variable used in statistics?
    (a) A variable usually corresponds to some measurable feature of the subject
    (b) A variable is a person, object, or event to which a measurement can be applied
    (c) A variable is the result of a particular measurement
    (d) A variable is the collection of values resulting from a group of measurements

13. The series "President, Vice-President, Secretary, Treasurer, Board Member" is on which type of scale?
    (a) Nominal
    (b) Ordinal
    (c) Interval
    (d) Ratio

14. Which of the following components of statistics contain error?
    (a) Measurement
    (b) Statistical analysis
    (c) Sampling
    (d) All of the above

15. If we have a set of measurements that are valid, but not very reliable, they will...
    (a) Be clustered around the right value, but in a wide cluster
    (b) Be clustered very closely together, but around the wrong answer
    (c) Be in a wide cluster around the wrong value
    (d) Include at least one measurement that is exactly the right value

16. Validity is how statisticians talk about minimizing _____ error; Reliability is how statisticians talk about minimizing _____ error.
    (a) Biased; biased
    (b) Biased; unbiased
    (c) Unbiased; biased
    (d) Unbiased; unbiased

17. When a comprehensive sample is not possible, what is the best sampling technique to use in order to avoid introducing additional bias?
    (a) Convenience sample
    (b) Stratified sample

(c) Random sample

(d) Systematic sample

18. Which of the following is the end product of the procedures used for calculating a statistical measure?
    (a) A single summary number that characterizes all of the values used in the calculation
    (b) A statistical technique
    (c) A range of numbers that characterize the population of interest
    (d) A valid and reliable measure

19. Every inferential statistic is _____, but most descriptive statistics are not _____.
    (a) Inferential; inferential
    (b) Inferential; descriptive
    (c) Descriptive; inferential
    (d) Descriptive; descriptive

20. The *Rule of Insufficient Reason*, which states that when we have no reason to think that one possibility is more or less likely than another we can assume that the two likelihoods are exactly the same, is related to which type of probability?
    (a) Classical probability
    (b) Frequentist probability
    (c) Subjective probability
    (d) We cannot calculate probability

21. Which of the following is the advantage to frequentist probability and their use of the concept of the *simple experiment* over classical probability?
    (a) Simple experiments work even when the underlying simple events are equally likely
    (b) Simple experiments work even when the underlying simple events are mutually exclusive
    (c) Simple experiments work even when the underlying simple events are not equally likely
    (d) There is no advantage of frequentist probability over classical probability

22. Which of the following statements illustrate the importance of statistical independence?
    (a) It is the minimum guarantee we can have that there is no causal relationship between two events
    (b) It is the maximum guarantee we can have that there is no causal relationship between two events
    (c) It is the minimum guarantee we can have that there is a causal relationship between two events
    (d) It is the maximum guarantee we can have that there is a causal relationship between two events

23. Which of the following is *not* a problem of introducing statistical dependencies between variables?
    (a) They introduce sampling errors
    (b) They introduce the problem of post-hoc hypotheses
    (c) They cloud our ability to spot real causal relationships
    (d) All of the above are problems

24. In a theoretical distribution, the x-axis represents the _____ and the y-axis represents the _____.
    (a) All possible values of the variable; the frequency of values in the data
    (b) All possible values of the variable; relative probability of each value
    (c) The range of values in the data; the frequency of values in the data
    (d) The range of values in the data; the relative probability of each value

25. If two bell curves differ in their central tendency, the curves will differ in which of the following ways?
    (a) The skew of the curves
    (b) The truncation of the curves
    (c) The position along the x-axis
    (d) The kurtosis of the curves

Preparing a Statistical Report

Part Two is written as a step-by-step guide to preparing a statistical report, from the initial idea for the statistical study, up through the presentation of the report. This organization is particularly useful if you are asked to prepare a statistical report as a homework assignment, or if you are preparing a report on the job.

Chapter 4 "What Is a Statistical Study?" defines a statistical study and the process of creating a statistical report. In Chapter 5 "Planning a Statistical Study" you will learn how to plan the study to get the results needed to support the business decision you are working on. Chapter 6 will guide you through "Getting the Data," whether you are gathering freely available statistics, buying data, or performing a study yourself. Chapter 7 introduces many varieties of "Graphs and Charts" you can use to convey information clearly. Chapter 8 introduces the most "Common Statistical Measures." Chapter 9 "Meaningful Statistics" addresses the questions of significance and meaning, allowing us to decide whether our statistical results support particular business decisions. And, in Chapter 10 "Reporting the Results," we learn how to prepare and present a statistical report.

Our practical organization, while it supports planning, preparing, and delivering statistical reports, deals with things in a somewhat different order than is found in most statistics textbooks. If you need to read these topics in the order taken by your textbook, please consult the web page for this book *www.qualitytechnology.com/books/bsd.htm*, where we maintain an up-to-date cross-reference to the most popular business statistics textbooks. Don't worry that you will be reading these chapters out of order. Now that you have the basics of Chapters 1, 2, and 3 down, you can read Part Two in any order that works for you.

One thing we will *not* be covering in Part Two are the more sophisticated statistical procedures, particularly those used to calculate inferential statistics. Statistical procedures are the subject of Part Three.

CHAPTER

What Is a Statistical Study?

At some point early in your business career, if it has not happened already, you will be confronted by a statistical report. It will not differ much from any other business report, except that it will be chock full of statistics in the form of graphs, figures, numbers, and tables. It may be large or small, readable or unreadable, fair or biased. The statistics reported will most likely be the results of a statistical study. A statistical study is just a project whose outputs are statistics. The statistical report makes those statistics available to an audience.

Here are the keys to making a good statistical report for business:

- *Focus on the business decision.* Make it useful, don't focus on the statistics.
- *Open with a one-page summary.* In the summary, identify the decision to be made, or the decision already made. Identify the questions asked, and their answers.

- *Put supporting material last.* In the body of the report, explain what you did and what it means. End with a conclusion guiding or supporting the decision. In this section, do not include any justifications or detailed explanations.
- *Save the details for the back of the report.* Whether you call this an appendix or additional chapters doesn't really matter. But put all supporting materials: your logic for choosing the statistical procedures you chose, your methods, your demonstrations of statistical significance, and your data, in the back of the report.

HANDY HINTS

Business is the Opposite of School

In school, you will learn what to put into a business report, but you may not learn how to write one. In fact, most school training and scientific training teaches you to do the exact opposite of what is needed for business. In school and science, we are taught to present our data first, our reasoning second, and our conclusions last. Also, the assumption is that we are not experts, that we have to prove ourselves.

In business, the opposite is true. The technical worker who prepares a report is assumed to be the expert. The business people want answers—your answers. Be the expert, and answer their questions up front. Reserve your reasons for when you are asked—if you ever are.

In short, do good work, present your results, and be able to explain or justify your work, but figure you won't have to.

The first key to a successful statistical report for business is the way the statistical work supports a business decision. Your first job should be to understand the questions being asked and the decision to be made. If you keep that focus and work through Part Two, you will be able to plan a simple statistical report, though you will probably want some help from an experienced statistician. If you are involved in a large or complicated statistical report, you can help by keeping the focus on the business needs, and coordinating the efforts of expert statisticians toward solving the business problem. Equally important, you will be able to review a statistical report, determining whether you believe the report's conclusions to be valid and to be relevant to the business decision it is intended to support.

Why Do a Study?

In a business context, there are three reasons for doing a statistical study.

- *Providing information in support of answers to questions that are likely to arise.* A business believes that it can anticipate certain questions that will arise, such that the data we collect will be useful in answering future questions. This type of study is usually done by companies that sell the results of their statistical studies, or inside large companies.
- *Providing general information about a situation, such as a business market, in order to define what questions should be asked.* In this case, a company realizes that gathering information is useful, but doesn't know exactly what information would be most useful. This kind of study is often performed by—or contracted out by—a small to medium-sized business that wants to learn more about its customers or employees. These studies often use standard questions that are generally accepted as being useful in business. These studies answer the question: What questions should we be asking?
- *Providing support to answer specific business questions and make specific business decisions.* The most common reason is that we have a decision (or decisions) to make, and we believe that general (statistical) information about the world can help us make that decision more effectively.

In any case, our first step is to determine what data need to be collected and what statistics need to be calculated. The best way to do this is to phrase a question or questions that, if we knew the answer, we would understand our subject, or our decision would be easier to make. That question will lead us to our data. The form of the question will help us decide what type of statistical measures or procedures are appropriate. These are the core elements that will drive our plan for the statistical study.

Why Use Statistics?

The executive functions within a business by making decisions that lead the company in a particular direction. Management functions by making and enacting decisions that improve the effectiveness and efficiency of moving in the right direction. If the executives and managers don't make good decisions, the company will head in the wrong direction or not move quickly enough.

Good decisions require good information. That is the basis of the business fields called *decision support* and *knowledge management*, and business statistics is part of those. There are only three kinds of information available for a business decision:

- *Hard facts*. If the quantity of information is small, we don't need statistics. For example, if a sales manager has names, corporate information, and degree of interest information about five hot prospects, and knows the sales team, the manager doesn't need statistics to make a good decision regarding which prospect goes to which salesperson, or what advice to give each salesperson to close the deal.

- *Data plus statistics*. When the information we need to evaluate is too large to handle all at once—that is, we are working with a large population—we turn to data sampling plus statistics. It replaces the complete picture that hard facts would give us in a smaller situation. And if the statistical work is good, the sample represents the population adequately, and our statistics tell us what we want to know, then the statistical study will support good decisions. This is what a sales manager would need, for instance, in deciding how to grow business in five different cities with tens of thousands of potential consumers in each city.

- *Assumptions and guesses*. If we don't get the hard facts or use statistics, we are left with assumptions and guesswork. All too often, business runs this way. For example: A small business hears that a nearby city is one of the fastest-growing in the country and decides to open a new store. Unfortunately, the business sells childrens' toys, and the city is the fastest-growing retirement community in the country. The store fails because of the assumption that a fast-growing city has lots of children. This is the kind of error that a good statistical study can prevent.

KEY POINT

When the dataset we need to look at is large, statistics is a lot better than assumptions and guesses.

Here is how we can make sure that we can use statistics well in support of a business decision:

- some general information about the world must be useful in helping make a business decision
- that general information must be calculable from specific information about the world that can be measured reliably and validly
- the specific information must be obtainable at a reasonable cost
- we must be able to present the results of the statistical work in a form useful to making the business decision
- the total cost of obtaining the statistical information must be lower than the estimated benefit of having the information when we make our decision

It is often not possible to determine, or even estimate, whether or not using statistics will net a benefit to our business for any given decision. Most often, we use statistics when standard business practices dictate that statistics have been of net value in the past. Sometimes, we need to use common sense and good judgment. If the cost of a statistical study is modest relative to the rest of our operations involving the decision in question, and the value of the information to that decision is clear and obvious, a statistical study is justified.

What Are the Key Steps in a Statistical Study?

The steps in a statistical study are:

- *Define the business decision to be supported, and the questions to be answered.* Most likely, you will be assigned to do the study by an executive or manager. This is just the beginning. You will need to meet with that person, and others, to make sure you have a thorough understanding of the issues, so that you can determine what data and what statistics will support the decision.
- *Create a research plan.* Creating a research plan is essential, and Chapter 5 "Planning a Statistical Study" will show you how. As part of the plan, you will decide what statistics support the business decision. This is a big challenge, and Chapter 9 "Meaningful Statistics" explains why statistical rules developed for science don't do so well for business, and what you can do about it.
- *Collect the data.* Chapter 6 "Getting the Data" will show you how.
- *Analyze the data* using the appropriate statistical procedures from Chapter 8 "Common Statistical Measures" and Part Three.

- *Report the results* choosing from the charts and graphs described in Chapter 7 "Graphs and Charts" and methods for written and oral presentation from Chapter 10 "Reporting the Results."

We must create the research plan before we jump in and get to work because all of the other steps are linked in a complicated way. Also the links do not run in just one direction. The decisions to be made can tell us what questions we want answered, but which questions can be answered depend upon the availability of the right data (or our ability to collect it) and applicable statistical techniques. We need to choose the right statistical procedures to answer our questions, and then collect the right data in the right way to use those procedures. But if our data sources are limited, we might not be able to. Then we have to rely on other, less optimal statistical techniques. All of this should be worked out in planning. If we work it out as we go along, costs will skyrocket as we are forced to do rework, perhaps even having to run a second survey or experiment.

During the planning stage, we consider all the issues of data and methods at once. In contrast, the statistical study itself must be conducted in a strict order. To avoid potential problems with post-hoc hypotheses and other problems of sampling discussed in Chapter 3 "What is Probability?" after we have our plan, we must follow this order:

- Collect the data
- Analyze the data
- Answer the questions
- Make the decisions (or allow our readers to make them)

Finally, when we write our statistical report, we must re-order things one more time, for clarity:

- Present a one-page summary of the decisions, questions, and answers
- Describe the decisions to be made
- State the questions
- Describe the data
- Report the analysis, showing the answers to the questions (or that we didn't get answers), and, if appropriate, recommending a direction on the decisions to be made, or explaining how the statistics supported the decision that was made.

In order to determine what data and what statistical techniques are needed for each question, we will need to understand the requirements and assumptions of the statistical procedures that will be used. This book, and our statistics training, will help us with that.

Planning a Study

There are two reasons why it is essential to plan a statistical study well.

- *The business reason* is that a well-planned study costs much less than a poorly planned—or unplanned—study

- *The statistical reason* is that, without careful planning, we can easily get the wrong data and be unable to run our statistical tests, or work in the wrong order, and invalidate the entire study by introducing bias

As in almost every facet of business, the key to an effective statistical study is good planning. Because of the costs and complexities of even moderately simple statistical studies, the value of planning is even greater. The real work of statistics is in planning what to do more than doing it. This is because most of the really dangerous and costly mistakes happen in the planning, not in the execution. In Chapter 5 "Planning a Statistical Study," we will present the planning process in order. Here are the most important issues.

KEY POINT

Everything that is true of a business report is also true of a statistical report. We need to do all the things that make a good business report, plus all of the things needed to make the statistics good as well.

Once we have our research questions listed, we need to make a practical assessment of whether or not we can realistically (and economically) obtain this information. This sort of practical assessment is not as easy as it seems. It requires thought and hard work. Lots of information that seems unobtainable may be sitting out there on a government website for free. Lots of information that seems straightforward and easy to obtain may be impossible to get. Only training and experience will teach us how to find out what sort of information is and is not available in our industry. And sources of information are changing all the time. The Internet has tons of information today that it didn't have yesterday, but much of this information is unreliable. Government restrictions and changing cultural norms make information that was easy to obtain 50 years ago almost impossible to obtain today.

The important thing to understand is that many statistical studies are begun without a clear understanding of the need to establish the costs and availability of the data required. Often, such studies are never completed, and large amounts of money and resources are wasted because no one took the time to find out if the data could be obtained at a reasonable cost.

Once a source or sources of data have been identified, a detailed plan for collecting the data must be constructed. The data collection plan must take into account the limitations and restrictions on how the data can be collected, whether due to costs, government regulations, or whatever. In addition, the data collection plan must use sound sampling procedures to ensure that the data are appropriate for whatever statistical techniques might be applied. Some of the restrictions on sound sampling apply to any sort of statistics we might do, and some of them are required only if we are going to do certain specific procedures.

A good example is *sample size*. Generally, the more information we collect, the better off we will be statistically. However, some statistical procedures require a great deal more data than others. We need to consult carefully with the project statistician before we make our data collection plan in detail.

What Are Data and Why Do We Need Them?

In Chapters 1, 2, and 3, we learned the theory of what data are. Here we will take a closer look at the practical side of that same question. Lots of information we use every day in business is data, even though we don't think about it that way. We need to be able to take advantage of opportunities to get high quality data at low cost. In order to do that, we need a better understanding of what data are, in practical, day-to-day business terms.

Collecting data for their own sake is a waste of time and money. If we want to apply statistics to our data, the data must be of the right type and quality. We need data only if we can calculate statistics from the data, and the statistical results will support the business decisions our audience is tasked to make. It is our responsibility to ensure that this chain, from population to data to statistical analysis to decisions support, is unbroken and carefully documented before we go forward with the study.

After we have determined that data can help with the business decisions our audience is tasked to make, we need to figure out what sort of data are needed. Some issues are always important to consider: Can the data be obtained? Can the data be specified precisely enough that they can be encoded symbolically or numerically? What form will the data have to be in, in order for the statistical analysis to be performed?

What we need here is a *measurement plan*. As with many sorts of plan, working backwards from what we need to how to get it is a good idea.

Given our research questions, we should be able to envision what the answer should look like. From that, we (or our consulting statistician) should be able to specify precisely not only what information we need, but precisely what form it must be in. We create the measurement plan last, and use it first. The measurement plan will tell us how to take the information from the world and put it into the precise numerical or symbolic form required by the statistical analysis.

The part of measurement where the information is converted into numbers or symbols is called *encoding*. Coding systems can be very elaborate, and we will consider only the simplest here in *Business Statistics Demystified*. The most important thing to understand here is that not all information can be encoded easily, and some cannot be encoded at all. Practically, our measurement plan will describe how we will encode the data so that they are useful for our decision. After we prepare our data collection plan, we may revise our measurement plan to make it work with the data collection process.

Once we have determined the type of data we need and what form we need them in, we need to ensure that we can obtain them in usable shape. In order to ensure that the information provided to us is valuable, we need to ensure that it is correct. In order to ensure that it is correct, we need to ensure that (a) it really comes from the world and not from somewhere else, and (b) it has been extracted from the world in reasonably good shape. (These two desiderata are more or less the notions of validity and reliability, discussed in Chapter 2 "What Is Statistics?") Whatever data collection procedures we ultimately choose, those procedures must be carefully planned and those plans must be carefully documented. We need to ensure two things: Are the measurements taken reliable? Are the data collection procedures designed and implemented so as to ensure the minimum possibility that error will be introduced from the point when measurements are taken up until the statistical analysis begins?

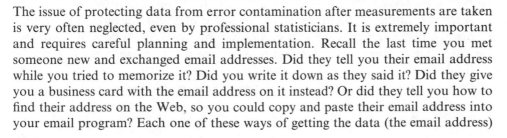

CRITICAL CAUTION

The issue of protecting data from error contamination after measurements are taken is very often neglected, even by professional statisticians. It is extremely important and requires careful planning and implementation. Recall the last time you met someone new and exchanged email addresses. Did they tell you their email address while you tried to memorize it? Did you write it down as they said it? Did they give you a business card with the email address on it instead? Or did they tell you how to find their address on the Web, so you could copy and paste their email address into your email program? Each one of these ways of getting the data (the email address) onto your computer has different problems and different possibilities of error. Each one, in order, is less likely to introduce error than the one before it.

The important thing is to apply common sense: Computers are more accurate than people; automatic transfer is better than copying; copying (or cutting and pasting) is better than retyping; typing is better than handwriting. Also, whenever possible, use two different methods of gathering data, rather than just one. If you can't do that, gather the data twice and compare them. Whenever possible, use both your eyes and your ears. If you can't use both, use your eyes as first choice. Make copies of everything, and verify everything you can.

The most practical output of measurement planning is a data collection form with instructions on how it is to be filled out.

Gathering Data: Where and How to Get Data

Once we have decided what type of data we need, we need to figure out the most effective way to get them. Usually, cost is the overriding concern. And the most important determinant of cost is the means by which we collect the data.

Assume that we know what kind of data we need, in what format, and that we have determined that, at least in principle, it is possible to measure the world in a way that will produce data in that form, we need to consider where and how to get the data. Our measurement plan shows us the form we want our data to be in. Our data collection plan will tell us where and how we will get the data to put into the form.

WHERE TO GET DATA: THEFT VERSUS HONEST TOIL

The first big issue in planning our data collection is: Do we already have the data we need somewhere in our own company? If not, has someone else already collected the data we need, and could we get the data from them? Tons of data have been collected and stored and made available for all sorts of purposes. Some of them are free and some of them we have to pay for. Before we begin the difficult and complex (and costly!) process of planning our own data collection and measurement, we need to see what is already available.

We need the data to be available at a reasonable price and also in a form we can use. Sometimes, someone else has collected the right data, but cannot, or will not, provide them to us in the right form. We may need daily records and the vendor may only sell monthly records, even if they originally measured things day-by-day and summarized them later. The data may have been encoded in a way that cannot be used for our planned statistical analysis, perhaps with inappropriate categorization or level of precision. Very often, the data may have been collected in a way perfectly appropriate for their original purpose, but useless for ours.

If data are out there in a form we can use, or in a form we can convert into something we can use, we need to establish that it will cost less to acquire the data from this source than it will be to collect them ourselves. It is usually cheaper to buy than to build, but not always.

HOW TO GET DATA: METHODS

If we decide to collect our own data, the big issue is what method to use. Very often, the method will be dictated by the nature of our research question, as we will see in Part Three. The two main methods are surveys and experiments. Quasi-experiments are a good compromise when practical limitations, such as cost, make experiments problematic.

Surveys

In a survey, we go out and gather data from the world, more or less taking it as we find it. The best known type of surveys are polls, where we contact various individuals and ask them for the information. The best known type of polls are opinion polls, where the information we ask for is the opinion of the person contacted. But actually, there are a number of types of surveys and polls:

- *An opinion poll* is where we survey people to get their opinions. We will probably also get some facts (such as their age) from them at the same time. Correlating demographic facts with opinions and preferences is important to marketing strategy.
- *A poll of people* can be used to obtain facts (as the sample perceives them) and preferences, as well as opinions. A good example is an exit poll, at election time.
- *A survey* does not have to ask people questions at all. When a surveyor measures land, he is doing a survey of the landscape, defining measurements that go onto a map (a data collection tool). In business, one common type of survey is the inventory, where we either count all items, or, for large inventories, use sampling and estimate inventory size and condition.

Experiments

Experiments differ from surveys and other non-experimental studies in that an intervention, called an *experimental manipulation*, is performed. The experimental manipulation is usually intended to model some change that might happen in the real world. This change is directly connected to our research question. For instance, if we wanted to know whether our new ingredient improved the flavor of a soft drink, we might add the new ingredient to some bottles of the soft drink and not to others.

In an experiment useful in a business context, the goal is to determine whether or not there is a correlation between the intervention and some desirable, or undesirable results. Ordinarily, we think of the intervention as causing the changed results, if any. Within the context of a properly designed experiment, statistics provides information that may lead to a conclusion about causation. Statistical calculations alone can only demonstrate the presence or absence of correlation, not causation.

When we conduct an experiment, we are looking to see if there is a change resulting from the intervention. Therefore, it is very common in experimental studies that the intervention is only done to some subjects or units. The subjects or units that are modified are part of what is called the *experimental group*, because they receive the experimental manipulation. The other subjects or units do not receive the intervention. They are part of the *control group*.

KEY POINT

The comparison between the results for the control group and the results for the experimental group will show us a difference that may have been caused by the intervention, if there is a difference. Without the control group, it would be hard—perhaps impossible—to demonstrate that the intervention was the cause of a particular effect.

Ideally, we want to choose which subjects or units receive the intervention on a random basis. In that way, the experimental and control groups are random (or at least quasi-random) sub-samples of our overall sample for our experimental study. Just as there are many ways to sample from our population, there are even more ways to divide our sample into groups who receive different interventions (or no intervention at all for the control group). This process is called *assignment* to groups and is a key feature to different kinds of experimental designs.

Quasi-experiments

Between surveys and experiments are *quasi-experiments*. *Quasi-Experimentation: Design and Analysis Issues* by Cook and Campbell (1979) is the best source of information on quasi-experimental studies. Quasi-experiments are used when proper control groups cannot be used or are too expensive (or unethical), but something more capable of finding possible causes is needed than a survey can provide. Quasi-experiments include interventions, but who gets what intervention (or any intervention) is not determined in the best way possible. Group assignment is performed in a non-random fashion.

There are many, many kinds of experiments and quasi-experiments. Each of them presents its own unique statistical challenges. In *Business Statistics Demystified*, we will focus on the most common experimental designs used in business.

Writing a Statistical Report for Business

Statistics can be intimidating. (You probably know that already.) While clarity is vital in any business report, it is much more important and much more difficult in a statistics report. As with any report, the first rule is to keep your target audience in mind. How often do they read this type of report? How sophisticated are they in terms of mathematics and statistics? Knowing how to present statistics with a minimum of numbers is a critical skill. Finally, remember the old seminary school adage: "Tell 'em what you're gonna tell 'em. Tell 'em. Tell 'em what you told 'em." Both the introduction and the conclusion should be expressed in straightforward ordinary language, laying out the business decision and relevant information.

Reading a Statistical Report

If you can plan and write a good statistical report, you also know most of what you need to be able to read and evaluate a statistical report. Why is it harder to read a statistical report than to write one? Because, here in *Business Statistics Demystified*, we are teaching you to write a good, honest, accurate, useful statistical report. But many of the reports you will read will be poorly done, biased, vague, or useless. And those problems may be hidden under layers of pristine prose and clear charts.

KEY POINT

One of the most common uses of a basic understanding of statistics for business is the ability to read a statistical report and answer the questions: Is this report any good? Is it relevant? Will it support the decisions we need to make?

The problem of bad statistics is not a new one. In fact, Huff and Geis wrote the best manual on reading statistics and seeing through bias in 1954, *How to Lie With Statistics*. We highly recommend this entertaining and educational little volume. In it, on pages 122–142, they encourage readers to watch for these misleading techniques found in statistical reports, and in articles and advertisements that present themselves as being based on statistics:

- *Who says so?* Does the corporate or individual source of the information imply a possibility of conscious or unconscious bias?
- *How does he know?* Examine issues of population, sample size, and sample bias. Sample bias can arise from intentionally biased selection, from poor design, or from self-selection.
- *What's missing?* Watch for an average that doesn't specify mean, median, or mode, and for conclusions not based on comparisons. If results are presented as an index, we must ask whether the index varied because of a change in the factor being discussed, or a change in the base of the index. And we have to pay attention to extraneous factors that may affect report results in such a way that the claims of correlation remain unproven.
- *Did somebody change the subject?* The link from data to statistics to conclusions may not be valid. For example, a survey may get people's self-report of their behavior, but that may be biased, and not represent what they actually do.
- *Does it make sense?* Sometimes, the underlying theory is invalid. In researching the rapid growth of the cellular telephone industry, the second author found numerous projections that indicated that, if current trends continued, everyone in the world would own five or ten cell phones in another decade or so. But that just isn't going to happen. And, if we go back in time, we will find out that this very kind of error was cautioned against in 1954, in *How to Lie with Statistics*.

Unfortunately, many of these misrepresentations and errors are hard to detect unless the survey data and research methods are available. If we have only the results of the study in hand, all we can do is be suspicious. If we can't acquire original data and a record of methods used, and we don't trust the source to be making a better effort than we could to be accurate and unbiased, we should gather our own data and do our own statistics. Otherwise, our company is making a business decision based on opinions and assumptions, disguised as a statistical report.

Quiz

1. A statistical study is . . .
   (a) A project that makes statistics available to an audience
   (b) A project whose outputs are statistics
   (c) A summary of the statistics
   (d) A document containing charts and graphs

2. What is *not* a key to making a good statistical report for business?
   (a) Focus on the business decision
   (b) Save the details for the back of the report
   (c) Put supporting materials first
   (d) Put supporting materials last

3. Which of these is *not* a type of information available for a business decision?
   (a) Hard facts
   (b) Data plus statistics
   (c) Assumptions and guesses
   (d) All of the above *are* types of information available for business decisions.

4. The first step in conducting a statistical study is . . .
   (a) Define the business decision to be supported, and the question to be answered
   (b) Create a research plan
   (c) Collect the data
   (d) Analyze the data

5. The _____ stage can be conducted in any order; the _____ stage must be conducted in a specific order.
   (a) Study; planning
   (b) Planning; study
   (c) Planning; report
   (d) Report; planning

6. The *business* reason for a well-planned statistical study is to reduce _____.
    (a) Error
    (b) Bias
    (c) Bad data
    (d) Costs

7. The process of converting information into numbers is ...
    (a) Sample size
    (b) Statistical analysis
    (c) Encoding
    (d) Data

8. Which is not a *method* for collecting data?
    (a) Survey
    (b) Experiment
    (c) Quasi-experiment
    (d) All of the above are methods

9. The _____ group receives the intervention; the _____ group does not receive the intervention.
    (a) Control; experimental
    (b) Experimental; control
    (c) Assignment; experimental
    (d) Experimental; assignment

10. What should you ask yourself when reading a statistical report?
    (a) Who says so?
    (b) What's missing?
    (c) Does it make sense?
    (d) All of the above

CHAPTER 5

Planning a Statistical Study

The need for planning in business is illustrated by *the 1:10:100 rule for Planning : Building : Using*. Missing a step in planning that would take one hour adds ten hours of work time to a project. If the work is missed during the project, it costs the customer a hundred hours of lost operations to fix the problem. For example, if an architect makes a mistake that he needs to fix, it would take him one hour of architectural work (planning) to fix it. But, if the error didn't get caught until after construction started, it would take ten hours (and cost ten times as much) for the construction team to fix it. And if the error was missed during construction, it would take 100 hours (with four days of lost rental income) to move everyone out and fix the problem once the building was in use. In construction, this is not surprising. What is amazing, but true, is that this rule applies in every project we do, whether we are building something, or gathering and analyzing data, or making a phone call, or preparing a report. It has been measured and proven over and over. The 1:10:100 rule is about as close to a universal law as you will get in

business. You can reduce the costs of your statistical study by a factor of ten by planning it well.

BIO BITES

The first author spent a brief time as an assistant account executive at a small midtown Manhattan advertising agency. I shared an office with three other junior executives. My boss had asked me to gather some information, via telephone, from five or ten companies. I had done such small, informal phone surveys before, so I felt confident. I stumbled through the first phone call so badly that my office mate, who was not a particularly nice guy, took pity on me. He sat me down and showed me how to plan a five-minute phone call.

I remember how shocked I was that something that small was worth planning for. My office mate had a nice, simple system that took no more than three minutes, using a pencil and one sheet of paper. The plan was no more than a list of topics, in order, that I needed to cover in the phone call. I use it to this day.

The lesson: No matter the size of the task, so long as the planning process takes less than ten or twenty percent of the time the task will take, it is worth taking the time to plan.

For a number of especially obvious reasons, statistics benefits more from planning than do most areas of business. In statistics, because of all the calculations and the need for both precision and accuracy, tiny mistakes can have catastrophic consequences. In addition, in business statistics, we have to present to an audience that doesn't know statistics. If we spend half our time planning, that is probably not too much, and may not be enough.

We will need a detailed research plan in order to conduct the research that will produce the contents of our statistical report. The structure of our plan will differ depending on whether we will be collecting our own data or obtaining pre-collected data from some source. In either event, these are the steps for creating a research plan:

- determine the plan objectives
- state the research questions
- assess the practicality of the study
- plan the data collection
- plan the data analysis
- plan the statistical report
- prepare a budget
- writing up the plan and getting it approved

This chapter will show you how to do this process from beginning to end.

Determining Plan Objectives

A statistical study begins with a directive, perhaps from your boss, that may be written or verbal. It could be as simple as your boss saying, "Find out about our customers in Chicago." Or it could be a full, clear list of questions, a sample report for how the results should look, and a date it is due. In all probability, the directive will not contain enough information for your plan. By meeting with your boss and others, try to get clear answers to all these questions:

- When is the report due?
- How much money and time are budgeted for the statistical study?
- What decision(s) will the information support? What is the issue, and what are the specific questions? Is there a specific act of planning, opportunity under consideration, or problem to be solved?
- What questions need to be answered?
- Is there a sample of a similar report or study, done in the past, or done by another company, that you could use as a model? If not, find some, and show them to your boss and have the boss define what is needed.
- Will the report be presented orally, in writing, or both?
- Are there any rules or restrictions governing how you do the study? For example, does your boss want you to use available data, or to do a survey or experiment, or is it left up to you?

In an ideal business, you would get all this information right up front. In reality, getting it may be like pulling teeth. If so, figure that you were asked to create the study, and be creative! Trust your judgment, come up with what you think is best, and then outline what you plan to do. Illustrate it with a similar report from another study, or a mock-up of the report you will create. Bring it to your boss, and get an okay, or, if your boss has some ideas, there may be changes. Once you have the business issues, format, budget, and deadline defined, you are ready to turn the business questions into statistical research questions.

Defining the Research Questions

Now is the time to bring in a statistician to help with the plan. Together, you can translate the business questions into a series of statistical issues that will define the study:

- What is our population?
- What is our sample size? How will we obtain our sample?
- Will we gather our own data, or use data from someone else?
- Will we generate our own statistics, or use statistical results from someone else?
- What statistical measures and procedures will give us information that will support our business decision?

There is no easy way to do all this, and it definitely requires a consulting statistician. As we explain in detail in Chapter 9 "Meaningful Statistics," statistics has been defined primarily to serve science, and the notion of statistical significance, in particular, was defined in ways useful to science. As a result, it is very difficult to translate business value and business significance into statistical significance. This is exactly what a statistician can help you do.

CRITICAL CAUTION

Many statistical consultants will charge more per hour to rescue a project than to plan for one, so we save money both in terms of hours spent and on the hourly rate.

As you work with the statistician, your job is to keep all the business requirements and constraints in mind. The statistician suggests methods of getting the data and the statistics, and statistical procedures that will provide useful results. The structure of the research questions will tell that statistician what statistical procedures are most useful. The content of the questions will help you define the population, the sample size, and where to get the data. Together, you seek a cost-effective way of creating a study that will provide statistical results that support the decision to be made.

For example, the decision to start a new advertising campaign requires a lot of information. Knowing the answer to the question: "Who will like our product?" might be helpful, but it is poorly phrased. A question like: "What demographic groups who are prone to purchase our product are currently unaware of it?" is better. A question such as: "Is this product more likely to be purchased by older or younger people?" is too general. Rephrase it as: "Which age groups, 8–18, 19–34, 34–50, 50 and over, are most likely to purchase our product?"

Turning a general or vague business question into a good research question requires the following:

- *Defining the population.* In the above example, are we talking about potential customers in the USA, or in just one city, or worldwide?
- *Defining the attributes of the population you want to measure.* Age is one attribute. Gender is another. What about household income level, buying for self or others, spending power, or ethnicity? All might be valuable.
- *Defining the required precision for each attribute.* For example, an age range may be just as useful as an age exact to the year.
- *Align all of these elements with business and statistical standards.* For example, certain age ranges are used by magazines in describing their readership. If we use a different set of age ranges, we may not be able to use our study to decide which magazines we should advertise in.

Even if we get all of this right, have we really supported a business decision? It's not clear that we have. If all we know is the demographics of people in the right market associated with likely interest in our product, we probably still don't know how to reach these people. This study, or another study, or expert advice, will be needed to determine whether we reach the target audience by print ads, television ads, posters in a supermarket, demonstrations at health expositions, household parties with personal networking, or whatever. Why? Because the real business question is not "Who buys our product?" That's not enough; we need to know, "How do we reach people who would want to buy our product?" and "What information, in what context, influences people who would want to buy our product to actually make the purchase?"

As we work to define the population, the sample, the questions, and measures, we also have to think about how we will get that sample, and how we will ask the questions and take the measurements. From the business side, the key issue is cost. From the statistical side, we have to address two other concerns:

- *Preventing bias.* In various situations, people will report consistently inaccurate results, or tell others what they think others want to hear. See Chapter 15 "Creating Surveys" for more about this. If we are using experiments or quasi-experiments, we will have to avoid bias as well. For this, take a look at Chapter 6 "Getting the Data" and consult with a statistician.
- *Ensuring data meet the requirements of the assumptions of the statistical procedures we will use.* Each statistical procedure has certain

assumptions which are, essentially, requirements about the data and certain statistical measures of the data.

Working with a statistician, and considering all of these factors, we build the plan to gather the data, analyze them, and prepare the report.

Assessing the Practicality of the Study

Once we have a clear set of questions, we know our population and sample size, and we have a rough idea of how we will get the data, we can do a cost and time estimate. We then compare our estimate with our budget and delivery date. Is it possible to do a useful statistical study with the time and money we have? If so, we move ahead to create a detailed plan. If not, it's time to tell the boss, and work out whether to drop the study, or to give it enough time and money to do it well.

Preparing the Data Collection Plan

Where and how will we get our data? Our first question is: Do we want opinions, or facts? We will get opinions from an opinion poll. We will get facts from a survey, an experiment, or a quasi-experiment. But, if we try to get facts from a survey of people, we need to be particularly careful about biased self-reporting. The question we need to ask is which of the following methods is the least expensive acceptable source for our data and statistics? Items are listed from what is typically least expensive to what is typically most expensive.

- Using data the company already own
- Acquiring (for free) or buying statistical results and, if available and affordable, the data that they are drawn from
- Acquiring (for free) or buying data, and generating our own statistical results
- Performing a survey
- Performing a quasi-experiment
- Performing an experiment

In all likelihood, the first option that is sufficient to the business purpose will be the least expensive. But, if the budget is available, consider more expensive options if they will give higher quality information and better decision support. If your company is making a major decision, then the cost of making the wrong decision may outweigh the added cost of a better statistical study.

Refer to Chapter 4 "What Is a Statistical Study?" for issues concerning the validity, reliability, and business relevance of these different sources of data and statistics. Later in this chapter, and in Chapter 6 "Getting the Data," we provide methods for planning and performing data acquisition. It is also very important to understand Chapter 9 "Meaningful Statistics" as you prepare your data collection plan. Before you finish the data collection plan, you and the statistician will have to choose measures and statistical procedures from Chapter 8 "Common Statistical Measures" and from Part Three and ensure that your data collection methods result in data that meet the requirements of the assumptions of the statistical procedures you will be using.

CRITICAL CAUTION

A Self-Description is an Opinion, Not a Fact

You may think that if you conduct a survey and ask a factual question, such as "How old are you?" that you would get a fact as the result. This is not true. When people report on themselves, you have only their answer to your question, which may not represent reality. For example, hospital emergency rooms track the age a person reports themselves to be, and also the age they are later shown to be through some form of verification. This is crucial to their business. In an emergency-room setting, a doctor may make a crucial decision based on the age the patient reported. Suppose that the patient—out of vanity, perhaps—claimed to be ten years younger than he actually was. The procedure led to serious complications due to the patient's age. Now, in a review or lawsuit, the hospital will have to know both how old the patient actually was to explain the medical events, and how old the patient said he was, to explain the doctor's decision. This inaccurate self-reporting happens frequently enough that hospitals maintain systems to track it.

In some cases, such as rating an employee's quality of work, we may have both objective facts and opinions. In other cases, such as measuring social or psychological values such as trust on a team, we have only opinions, and no definable facts. In addition, we have difficulty being sure that everyone we survey has similar understanding of the meaning of the term. This is a major problem in psychological, sociological, and business surveys.

OTHER PEOPLE'S DATA

Most often, the least expensive data are data someone else has already collected. The US government collects a lot of valuable business data and makes it available for free. In addition, many companies are in the business of collecting data and selling them to businesses. It is always easiest to find data that someone wants to sell you, or to give you.

In addition, the data we want may have been collected in the ordinary course of business by our customers, our vendors, or even our competitors. In this case, the data are likely to be proprietary, and we may not be able to get ahold of it. Before we abandon this alternative, however, it cannot hurt to ask. Just recently, the second author of this book wanted some information held by a client, which he was sure was proprietary. He asked, and the vendor was happy to provide it. Very often, companies are willing to share information they have, if only for a price. This is especially true if you find a business that is in your market, but is not a direct competitor.

Finding data already collected is much like finding anything else. A web search is a good start. There are also books that catalogue various sources of data. You can get some excellent information about many places to look for data, either for sale or for free, from the authors' web site, *www.quality technology.com/books/bsd.htm.*

Planning Data Analysis

The best way to plan the data analysis is to use a backward pass, followed by a forward pass. Start with the business decision, and then go to what you hope to show in the statistical report. From that, working with the statistician, determine which statistical procedures you want to apply to your data. Then plan to collect the data in a way that will allow you to do those procedures.

Now, do the forward pass. With these data, as collected, do you meet the requirements of the assumptions of the statistical procedures you want to use? Can you modify data collection procedures, or the data collection form, to make it easier to do the statistical work? Then define what computer programs and other tools you will use to generate and check the statistics. If appropriate, also write up the step-by-step procedures. Then define what tables, graphs, and charts will display the data in the statistical report. Can you derive all these report items from the statistical procedures you are using? Will they give the business what it needs to make a good decision?

Along the way, you may have to make trade-offs. For example, data meeting the requirements of a certain procedure may be too expensive to collect. In that case, you drop the procedure, and see if you can find an alternate way of supporting the business decision. When you are done, you should have an affordable process of collecting data, analyzing them, and preparing the report.

Planning the Preparation of the Statistical Report

You can build a mock-up of your final report using these resources:

- The beginning of Chapter 4 "What Is a Statistical Study?," where we discuss the focus and features of a good statistical report for business.
- A report used by your business, or seen by your boss or customer, that the boss or customer likes.
- Other books on writing good business reports and delivering business presentations. One good source is *Budgeting for Managers*, by Sid Kemp, from McGraw-Hill. Chapters 6 and 7 discuss how to check a report for errors and prepare for a presentation.
- Chapter 7 "Graphs and Charts," which will let you choose the best charts and graphs to present your ideas.

Now, let's turn our attention to the fact that we are throwing statistics at a bunch of business people. Consider the audience, and consider their goal. They want to make a business decision. The statistical report is coming in as input, and probably not to be evaluated. Therefore, proof that it is a good report is not an issue. A clear presentation of results that they can use in making a decision is what matters. Of course, if your audience is familiar with statistics, you have more leeway. And if some of your audience might have reason to challenge your work, you need to be prepared to respond to questions. But the focus should be on the business at hand, and the presentation, in length and format, should fit the situation.

Some things that are considered good statistics can be bad business presentations. For example, consider precision. In statistics, engineering, and science, precision is considered very valuable. But often it is not needed for business. It is a common fallacy to think that numbers and technical terminology are needed in order to be correct and accurate. Numbers only lend *precision* to a presentation. Numbers can't add clarity, and can often detract from it. Remember on *Star Trek*, when the Captain asked Mr. Spock how long before they arrived at the planet and Mr. Spock said something like, "Three days, 4 hours, 8 minutes, and 13.436 seconds." How useful was that information about the last .036 seconds? This is overprecision, time and resources taken conveying useless information. In addition to being only as precise as we need to be, we might consider that the most useful answer is not presented in numbers at all. Mr. Spock could have said, "We'll get there Thursday afternoon,"

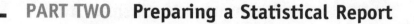

a much more digestible answer for a human audience. Similarly, the best answer to the question, "How much money will we make if we do this?" might simply be, "A lot more than if we don't."

CRITICAL CAUTION

Don't Assume Your Results will be Useful Until the Study is Done
In all of this planning, it is easy to make the mistake of assuming that you will be able to get useful results. Even if your planning, data collection, and analysis are excellent, it may simply be the case that the data you gather do not provide useful information. Suppose you are supporting a decision of which city of two cities to expand into with a new store. You gather statistical attributes of your customers. You gather data on the two cities. All along, you are assuming that, when you are done, there will be a significant difference between the potential customers in the two cities.

And there just isn't. The best that the statistical study can tell the decisionmaker is: "As far as statistics can tell us, both cities are equally good opportunities." Then that is just what your report should say. Even if that isn't what you had hoped and planned.

Writing Up the Plan

Once you have worked out all of the details of the statistical study, you should write up a plan. The first page should be a summary with these headings:

- *Purpose*. State the decision you will support and the questions you will answer.
- *Process*. Describe, in one or two sentences each, how you will gather the data, do the statistical analysis, and prepare the report.
- *Report*. Describe the length of the report, the number of charts and graphs, and whether you will deliver an oral presentation, a PowerPoint slide show, or something else.
- *Delivery date*. When will you deliver the report? If you are not starting the study right away, when will you start?
- *Cost*. What is the total cost?

That page is for your boss, and you will use it to get the plan approved. The rest of the plan is for you, the statistician, and the team working on the study. Using the same headings as in the summary, write up a detailed, step-by-step procedure of how you will prepare the report. Include data collection forms and anything else that will make the job easy. You might even build the data tables for your final graphs and charts, put in dummy numbers, and get everything looking nice. That way, if you are rushed at the end, you can drop in the data, run the statistics, drop in the results, and the report will be nearly done. Of course, you should be very careful that you adjust the charts and graphs to match the actual data before you deliver the report.

A large statistical study is a major project. For major projects, best practices should be used for time and cost estimation, and also for risk and quality planning. This topic goes beyond this book, but we can recommend another text in the series, *Project Management Demystified,* by Sid Kemp.

PREPARING A BUDGET

Your plan needs to contain a budget, an estimate of the money you will need to spend to complete the statistical study. Depending on your company, you may or may not count work hours for permanent staff in the budget. Even if that is not part of the budget, you should do what is called a time budget, so that you can get approval for in-house staff to work on this project, instead of something else that they might be assigned to do. In preparing the budget, consider these four elements:

- Internal staff time
- Consulting fees, either a fixed fee, or per hour with an estimate. If you get a per-hour rate, you may want to request a guaranteed maximum.
- Items to be purchased, if any
- Tools, such as computer software or #2 pencils, to be purchased, if any

To prepare the budget, simply estimate the time and purchase costs for each work item in your plan. For more detailed explanations of how to prepare time estimates and budgets, see either *Project Management Demystified* or *Budgeting for Managers,* both by Sid Kemp.

GETTING THE PLAN APPROVED

Sometimes, the plan is automatically approved when you are assigned the work of creating the statistical report. Other times, either the budget or the plan needs review and approval. In either case, you should present the plan and the budget, and get an okay to go ahead. That approval may be formal or informal, depending on how your company works and how much the statistical study is going to cost.

As you go for approval, there are some extra items to consider. We have guided you in preparing the plan as if your only goal is to provide a report supporting one set of business decisions. And that may well be best. But you should also look to the future, and ask if you can create additional value while you are doing this statistical study.

- *Making the study repeatable.* Is it likely that the business will want this information again in the future, perhaps even on a regular schedule, such as quarterly or annually? If so, you can plan things that will make the study easy to repeat, and show the value of reducing the cost of repeating the study in the future.
- *Gathering all the data that you can afford.* If you are doing a survey, a quasi-experiment, or an experiment, you will probably find that doing it at all is expensive, but getting more data while you do it does not add much to the cost. Getting more data than you think you need has three advantages. If the statistician decides that additional statistical procedures are useful for this study, you are more likely to be able to do them. If the business audience thinks of additional questions when they see your report, you may be able to answer them. And if another business audience, such as another department, could use data from the same population, they may be willing to pitch in and pay for part of the study.
- *Making the study a model.* Even if this exact study will not be repeated on this population in the future, similar studies might be useful for your audience. For example, if this study examines a few new cities to determine if they are viable markets for your company, you may want to set up the plan so that a similar study can be done in different cities in the future.

As you go for approval, you will probably have a sense of whether what you are proposing is about what was expected, or if you discovered that things are going to cost a lot more than your boss would like. If cost seems like it might be a problem, prepare options. For example, you can show the possibility of purchasing data, rather than collecting them, but point out that the data are somewhat out of date and also don't allow you to run some of the statistical procedures that you want. There is a possibility that all your planning will lead to a decision that the study is too expensive, that it is not worth doing. That is not a bad thing. By doing careful planning, you have saved the company from the mistake of spending too much money.

But we should end on a happy note. In all likelihood, your study will be approved. Why? Because the cost of a statistical study is a lot less than the cost of running a business on guesswork. You are adding real value to the company by preparing this study. And you have planned it well enough that it is likely to be an excellent study, and easy to carry through.

Quiz

1. The *1:10:100 rule* demonstrates the importance of . . .
    (a) Writing clearly
    (b) Conducting good statistical analysis
    (c) Planning
    (d) Fixing problems after they occur

2. What information do you need to determine the Plan Objectives?
    (a) When is the report due?
    (b) What decisions will the information support?
    (c) What is the budget?
    (d) All of the above

3. Working with a consulting statistician is critical to which phase of planning?
    (a) Stating the research questions
    (b) Writing up the plan and getting it approved
    (c) Determining the plan objectives
    (d) Planning the statistical report

4. Comparing the cost and time estimates to our budget and delivery date help us determine . . .
    (a) The research plan
    (b) Stating the research questions
    (c) The practicality of the study
    (d) Planning the statistical report

5. The most expensive method of data collection is . . .
    (a) Using data the company already owns
    (b) Performing a survey
    (c) Performing a quasi-experiment
    (d) Performing an experiment

6. The best way to plan the data analysis is to use a _____ pass, followed by a _____ pass.
    (a) Backward; backward
    (b) Backward; forward
    (c) Top-down; bottom-up
    (d) Bottom-up; top-down

7. The most important aspect of planning a statistical report is to . . .
    (a) Be as precise as possible
    (b) Look at a report from your company
    (c) Consider the audience
    (d) Use as many numbers as possible

8. The first page of the written plan should contain . . .
    (a) The purpose of the report
    (b) The process you will use
    (c) The cost of the study
    (d) All of the above

9. Your budget should contain what information?
    (a) Costs of man hours and materials to be purchased
    (b) Costs of man hours only
    (c) Costs of materials to be purchased only
    (d) Costs of preparing the report

10. How can you create additional value of your statistical study?
    (a) Make the study repeatable
    (b) Gather all the data that you can afford
    (c) Make the study a model
    (d) All of the above

CHAPTER 6

Getting the Data

Quality data collection means obtaining reliable, valid data at low cost that is suitable for our planned statistical analyses. Happily, a properly planned data collection gives not only higher quality, but usually lower cost as well. There are three things we need to plan:

- We need to identify the best *source* for our data.
- We need to choose *methods* that ensure reliable and valid data.
- We need to choose a research *design* that ensures the suitability of our data for our planned analysis.

This chapter will be devoted mostly to considering various sources for our data. We will also discuss proper data collection methods. Research design will be addressed in Part Three.

Stealing Statistics: Pros and Cons

When data are available, either for free or for sale, the data providers usually also provide summary statistics describing the data. If the summary statistics they provide are suitable for our needs, it will save us time on the analysis.

Some statistics can be calculated from other statistics, even without the detail data. Sometimes, we can just use the summary statistics provided directly.

A big advantage to using someone else's statistical results is that sometimes the detail data cannot be supplied, for security or privacy reasons. Medical data are a very good example. It is perfectly legal to provide information as to how many American citizens have AIDS, including breakdowns by state and ethnicity. It would be a very different matter to give out their names and addresses.

The two biggest problems in using someone else's statistics are that we cannot double-check the calculations done by the provider and we cannot calculate any statistical results the provider left out. Also, if the provider made any errors, we will inherit those errors and they will contaminate any further analyses we do using those statistics. If the provider failed to calculate particular statistics we need, there is no way to calculate those statistics without the detail data.

CRITICAL CAUTION

Inheriting Errors

Inheriting the errors of others is far more common—and far more costly—than you might think. One study traced small factual errors that got into textbooks and found that they stayed for decades, through several generations of authors.

FREE STATISTICS

There are free statistics everywhere. Many are worth just what we pay for them. Others are of excellent quality. Governments and non-governmental organizations (NGOs) usually provide free statistics as a public service. Other sorts of organizations often provide statistics for their own purposes. It is important to make sure we understand the provider's motivations. Even a well-intentioned provider with a particular agenda may be unintentionally swayed to provide inaccurate information. Without access to the original detail data, we must be able to trust the provider. That is, we must believe that the provider had at least as strong a motive and ability to provide reliable, valid data as we have.

Most free statistics are provided without much guarantee of their reliability or validity. The news media and public relations organizations and private firms and professional and industry associations deluge us with statistics every minute, whether we like it or not. These statistics should be used with great care, or not at all.

Often, free statistics are provided along with the detailed data as a convenience. Very large datasets are often accompanied by smaller datasets containing summary statistics. If we have any doubts about the statistics, we can double-check some or all of the statistics by calculating them ourselves from the detail data. If we plan to use the statistical results directly, it is always a good idea to check a random sample of the summary statistics against the detail data in order to make sure that the provider's calculations are sufficiently free from error.

NOT-SO-FREE STATISTICS

When we pay for statistics, we have a right to expect a certain guarantee of quality. If the original data are available, and we can afford it, we should buy it along with the statistical results calculated from the data. If the data are not available, the first question we must ask ourselves is, why not?

If the detail data are not available for reasons of privacy or security, we have a right to expect that the statistics we purchase are calculated at as detailed a level as possible. One example of this is US census data. Although census data are free, it is an excellent example of how summary statistics can be provided at many levels of detail, while still maintaining the security of the data and the privacy of the persons surveyed. The detail data are secure by law. You will not get names and addresses from the U.S. Census Bureau. Summary statistics are only provided down to a level of detail where private information about citizens cannot be calculated in reverse. (For instance, if there is only one Hindu family in a particular area and we are told that the average family size for Hindu families in that area is five, then we know that that one family has five members.) The Census Bureau takes great pains to ensure that such reverse calculations cannot be done.

The Census Bureau demonstrates that data can be provided without endangering security or privacy. Therefore, if someone offers to sell us summary statistics without the detail data to back it up, we need to inquire carefully as to why.

Someone Else's Data: Pros and Cons

In general, obtaining detail data from an outside source is preferable to obtaining only summary statistics. The cost of computer power is so low, and the ease with which most summary statistics can be calculated using spreadsheets or simple statistical software is so great, that the benefits of having a vendor calculate summary statistics for us are usually negligible. The only issue that usually arises is whether it is better to use someone else's data or collect them ourselves. Other people's data are usually cheaper, but we do not have control over how it was collected, so we need to worry about quality.

The data collection procedures used by the vendor need to be documented in detail. If the vendor has a good reputation and our statistician informs us that the methods used by the vendor to collect the data make them suitable for our planned statistical analyses, then acquiring the vendor's data is probably preferable to collecting our own.

FREE DATA

When the data are free, the only issues are quality and suitability. As with free statistics, we need to be sure that the provider of free data does not have an agenda—conscious or unconscious—that might cause bias in the data. Having the data at a detail level limits the sources of bias, but unreliability and invalidity can enter the picture from three causes: conscious bias, unconscious bias, or poor data collection methods. The best insurance we can have of the quality of the data is detailed documentation of exactly how the data were collected. If that documentation is certified by an independent agency, so much the better. Documentation as to the data collection method will also allow our statistician to determine if the data are suitable for our planned analyses.

Archived data

Most commonly, data we acquire were collected formally by the vendor, using standard methods, whether survey or experimental, or some other. These data are then archived and we can purchase copies from the archive. Of course, with archived data, we have to ask: Is it out of date for our purposes?

Non-archived data

There is another source of free data. Lots of data are collected for reasons other than statistical studies. Sometimes, these data can be obtained with proper concern for privacy and security. In a sense, these non-archival data are halfway between obtaining archived data and collecting the data ourselves. Finding usable non-archival data requires some creativity, since the data were collected for some other purpose.

CASE STUDY

A Data Shake-Up

A wonderful example of using non-archival data happened a few years ago in Japan. Japan is subject to earthquakes, many of which are severe and dangerous. Recent research in seismology suggested that detailed information as to how the earth moves during an earthquake may be very useful in determining future risk and in developing safety measures. A very clever seismologist realized that the information he needed was being collected throughout the urban areas of Japan 24 hours a day. Security cameras in stores provided moving images from which the seismologist could collect detailed data about movements of the earth. The cameras were distributed across the area. All he had to do was to acquire the data after an earthquake from the area where the quake had occurred. The tapes were normally erased and re-recorded only after 24 hours. By notifying hundreds of stores throughout the city in advance, and arranging for the tapes to be saved after an earthquake, the seismologist was able to obtain more than enough data the next time a serious earthquake hit the area.

NOT-SO-FREE DATA

Paid data are to free data as paid statistics are to free statistics. We need to worry less about bias, since the vendor's motive is usually profit. However, we need to worry more about quality. We do not want to purchase data from a vendor who has skimped on data collection procedures in order to increase profits. Poor methods make for data that are not a bargain at any price. Once again, thorough documentation of the data collection methods, preferably audited by an independent agency, is the best solution. The reputation of the vendor is also important.

Doing it Yourself: Pros and Cons

The most common reason to collect our own data is if suitable data are not available from other sources. Data collection is almost always expensive. Collecting data using quality methods is more expensive. Failing to use quality methods is the most expensive of all, because the errors in the data will eventually cost us more than doing it right the first time.

The main advantage to collecting our own data is that we can have complete control over the data collection process. Proper detailed planning in consultation with our statistician is our best guarantee of quality data at low cost. In addition, collecting our own data means we can collect exactly the data we need using the methods most suitable to our statistical analyses.

DATA YOU ALREADY HAVE

Before planning an expensive data collection operation, we need to make sure that we don't already have the data we need. Many questions can be answered with data from within our own operation, even without conducting a research study. The answers to some of our questions can be found in company records, accounting records, personnel records, financial records, inventories, computer event logs, and business records of all types. This information is free and, because we don't have to take it outside the company, it can stay secure. Even so, planning for security is critical. Whenever data are moved from their home computer, even within the company, security can be compromised. A detailed plan for keeping data secure during every transfer and at every location is vital.

If the data available, either within the firm or outside, are inappropriate to answering our research questions, we will have to collect data, either by survey, experiment, or quasi-experimental study.

Here are some examples of data that are readily available inside most companies:

- *Financial records.* The detailed data unit of a financial record is the transaction, which records the date, the account that money moved from, the account it moved to, and the amount of money involved. The general ledger is the record of all transactions, and statistical analysis of these transactions can often show us things that standard financial analysis would miss.
- *Human resources (HR) records.* Stripped of names to protect confidentiality, records from HR can be the basis of statistical studies providing crucial input on key business issues.
- *Timesheets.* Timesheets are only accurate records if they are produced during the day, or, at latest, at the end of each workday. Studies have shown that, no matter how well intentioned, a timesheet prepared at the end of the week is essentially an act of fiction. Accurate timesheets, however, become the basis of statistical studies of productivity.
- *Manufacturing process records.* In the field of total quality management (TQM), data have been collected about manufacturing processes, both from totally automated production and from assembly lines, and then used to improve the processes.

- *Computer data logs*. Most computer systems have built-in methods of *logging*, that is, capturing computer transactions and events and recording them into a log file.

CASE STUDY

Sometimes, the Data Lead to the Survey

So far, we've assumed that a business decision, or a set of business questions, instigates the statistical study. That is not always the case. The second author of this text had a business client come to him and say, "We're getting all this data about how our customers use our web site. What can we do with it?" In this case, the proliferation of available data led to an astute entrepreneur seeing an opportunity. We evaluated the data and statistics the site provided, and worked with him to prioritize the business questions we could answer.

COLLECTING DATA FROM YOUR OWN OPERATION

If the data we need have not already been collected, often the simplest and least expensive method of data collection is to alter our usual business procedures in order to collect data on our ongoing operations. Sometimes, this can be as simple as turning on an automatic log on a computer, or purchasing and installing data collection equipment or software.

In other situations, we may need to change manual procedures. We may want managers or workers to record their activities. If we are going to do this, we plan and implement with attention to these issues:

- Design the data collection form so that it gives us the data we need, in the form we need, to apply the statistical procedures we want to use.
- Keep data collection as simple and automatic as possible. For example, it is easier, and more accurate, to gather productivity data for workers using computers from the computers than it is to ask workers to fill in timesheets.
- Follow the rules for preventing error and data contamination from Chapter 4 "What Is a Statistical Study?."
- Be aware of concern—founded or unfounded—from managers and workers about privacy, misinterpretation or misuse of the data, job security, and micromanagement. Many people feel that holding onto their work methods is a form of job security, and sometimes they are

right. You may need to consult with an HR expert to work through these issues.

- Ensure the confidentiality and security of the data.

If we take care of these concerns, we can create an environment where the business generates the data it needs as part of ongoing operations. This radically reduces the cost of future statistical studies, and provides the additional value from having reliable data collected using consistent methods over a long period of time.

CLEANING UP OUR DATA

Every business faces problems—often major problems—in cleaning up data so that they can be used for various purposes. Here are a number of things to look at when we evaluate how we will clean up the data collected from our business:

- How do we find and correct recording errors?
- How do we handle missing records?
- How do we handle records that are in the data, but do not represent our population?
- How do we determine if elimination of erroneous records biases our sample? For example, if the data were collected by people at different stores, and one person consistently made errors, then we may have no good sample from that store.
- Once we have a set of valid data, we have to consider that to be a sample and evaluate its relationship to the population we want to examine.
- We then need to organize the data to meet the input requirements of our statistical analyses.

When using existing data from our company, we need to make a plan, and budget time and money to the work of evaluating and cleaning up the data.

Survey Data

By *survey*, we mean any non-experimental method for collecting data, that is, data collection without an intervention. When a survey involves contacting individual people and asking them questions (as opposed to a survey of Websites, or public records, or weather conditions, or land, or anything else), we call the survey a *poll*. When the questions used in the poll ask people their opinions, we call that poll an *opinion poll*. Chapter 15 "Creating Surveys" introduces the topic of preparing surveys and analyzing survey data. If your area of interest is marketing or human resources, you might well want to take an entire course on preparing and using surveys.

The most important issue in conducting a survey is to define our population and ensure a representative sample, as we discussed in Chapter 2 "What Is Statistics?." Once a good sampling technique has been planned, the issue in terms of data collection specifically is to measure things in a way that avoids bias. This is especially difficult in a poll, where we must develop specific questions that get the information we want from the folks we interview. Good question design is the most difficult challenge in polling, and we address it in Chapter 15.

The other issues related to data collection in surveys are more general, applying not only to customer surveys and polls, but to surveys of all kinds. We will address these issues here:

- redundancy, validation, and calibration
- storage
- database design

REDUNDANCY, VALIDATION, AND CALIBRATION

When we are not conducting a poll, it is very often the case that our measurements involve converting information from one form into another. For instance, we may need information that has been stored in non-computerized records. The most common way to extract the information we need is to have people survey the situation and enter what they find onto a computer. This process, called *data entry*, is one of the most error-prone processes in all of statistics. The only method for ensuring reliability is to have two or more people enter every single piece of data. This procedure is called *data validation*. Obviously, data validation doubles the data collection cost, at a very minimum. However, this doubling of cost is far, far less expensive than the disasters that can befall us if we do not validate.

We also need to minimize the number of times the information is transformed and/or transported. Every time data are copied or moved or translated from one form into another, there is a potential for additional error. Once data have been safely and reliably collected, the less that is done to disturb them, the better.

Another issue is whether automatic computerized translations and copying of data are "better" than doing things by hand. When the procedure is just a matter of copying, then automatic means, whether mechanical or computer-based, are much better than doing things by hand.

On the other hand, using the computer is not always the best answer. It is not so much that computers make fewer errors than people, but that they make different kinds of errors. When the risk of computer-type error is high, we should use people. When the risk of human error is high, we should use the computer.

SURVIVAL STRATEGIES

The problem with people doing data entry is called garbage in, garbage out. On the other hand, the problem with automated data collection is summed up in the adage: To err is human, but to really foul things up takes a computer.

The lesson: We are responsible for ensuring the accurate input and translation of data, whether by computer, by machine, or by people.

The most reliable form of data entry is probably the electromechanical device, where a piece of mechanical equipment, such as a thermometer with a bimetallic strip or an infrared detector, measures temperature or the length of products coming off the assembly line, or some other physical attribute. The measurement is translated into an electrical impulse and recorded on computer-readable media. Even these systems are prone to error, but that is a problem for engineers, not statisticians. Even in these situations, redundancy, either by repeated entry of the same data, plus comparisons, or by entry and checking of known data, is the best method of calibration for the elimination of bias and error.

We can use computer input devices, such as optical scanners, to input survey data in a similar fashion. Or we can design a computer interface, such as a web page, where our population enters the data directly. These methods can be highly reliable, but it is important to realize that the reliability does not happen automatically, or by default. Computer interface design and testing is an engineering discipline and art in its own right, and we should make sure that we work with experts who understand bias and its sources, redundancy, testing, calibration, and error correction.

If people are entering the data, there are several possible systems, including manual recording with later transcription, standardized forms to be scanned, or direct data entry.

Each of these has advantages and disadvantages. It may seem that direct data entry is best. But consider this: suppose the survey workers are under pressure to meet an impossible deadline. They might stay up late, just punching numbers into the computer—making up data. If they filled out paper forms, there would be physical evidence of this, such as the pattern of handwriting. The computer data entry leaves no physical trace. So, we would need to devise other means, such as a hidden time-stamp on each survey, showing when it was done and how long it took, perhaps tied to the phone system used for the survey, to detect such sources of error.

STORAGE

If we design a convenient, secure data storage system that will allow for easy input, secure storage, storage for all the types of data and file formats we need, an easy, appropriate retrieval that does not compromise security, we will need to copy and transfer the data fewer times, and we will have fewer errors. The data should also be backed up and archived appropriately, with proper security on the archives. Good security is a balance of security plus appropriate access. We should review all of these issues with the appropriate data systems manager, because the storage, encryption, and security requirements for a statistical study, especially one with HR or other sensitive data, are different than the requirements for storage of ordinary business data.

CRITICAL CAUTION

Change of Data Systems can Prevent Data Retrieval
Sometimes, we need to retrieve old data onto a new computer. Due to the rapid change of computer technology, this can be very difficult. When archiving data for long-term storage (which could be as little as a year or two), we should work to ensure that the physical medium of the storage is something that we will be able to read on a computer that we will have in two, five, or twenty years. The second author remembers magnetic tape, $8''$ floppy disks, $5\frac{1}{4}''$ floppy disks, and numerous other media that have come and gone, leaving data irretrievable.

DATABASE DESIGN

Just as we must plan the type of physical storage and location, we must also plan for the logical form in which the data are stored. There are many ways to ensure the safety and accessibility of our data, and we need to use them. Data should be encoded in a way that will allow either us or the computer to detect any errors that may occur during storage. The technology of error detection and error correction is far beyond the scope of this book, but we need to know that our system has such features and that they are turned on and working. If data need to be analyzed in a different form than that in which it was collected, both forms must be stored. When data are cleaned or edited to correct errors, the earlier "dirtier" versions must be archived in order to keep an audit trail. Redundancy is vital in data storage as well as in data collection. In addition, the audit trail should include an appropriate record of the processes used to clean up data, and it may be possible to link those process records to the data records in a well-designed database.

The single most important aspect of quality database design is never to throw any data away. Data should be recorded as they are observed and/or collected. Never, ever summarize data before recording them in the database.

CASE STUDY

The Case of the Sinister Statistician

The first author worked for a while assisting graduate students in doing the statistics for their research. One Master's student decided to give all her subjects a test that had ten or twenty questions. The score for each test was only three numbers, calculated from the questions. The student decided to score each test by hand and then record only the scores and not the individual answers to the individual questions. The first author practically got down on his knees and begged the student to enter all the answers and use the computer to calculate the scores.

The student explained her logic: She would have to score each test anyway. The scoring procedure was very easy to do by hand. By doing it this way, she only had to type in three numbers for each subject, rather than ten or twenty! (The study had fewer than 30 subjects.) She had already gotten her statistical analysis plan approved by her Master's committee and was not required to do any analysis of the original numbers, only on the three scores.

The first author (still on his knees, now practically weeping) pointed out that: Scoring by hand tremendously increased the possibility of calculation error. There weren't that many subjects. The author knew many tricks to make data entry easier and less prone to error. The author was willing to help. There would only be one copy of the original data (on the paper test forms) and it might get lost. What if the committee changed its mind?

The student, now entirely convinced of the author's sinister intent, proceeded to score the tests by hand. After her preliminary oral examination, one of the faculty members on the committee decided that a more detailed statistical analysis was needed based upon the individual test questions. The student had to return to the lab and spend half the night entering data in order to graduate that semester. By her own estimate, she more than doubled her own workload by scoring the tests by hand.

The moral to the story is this: The authors of *Business Statistics Demystified* are not sinister. We are not trying to make things harder on anyone. We have learned from years of our own experience and our own mistakes how to lessen the likelihood of making extra work for ourselves. It all boils down to this: Planning pays off, and it is better to get it right the first time. We are just trying to pass that information along.

Experimental and Quasi-Experimental Data

The first and most important thing to know is that all of the rules that apply to collecting survey data (above) also apply to data derived from experiments or quasi-experiments. As we will see in Part Three, the biggest difference between collecting survey data and collecting experimental or quasi-experimental data is that, in the latter case, we collect data sampled from more than one population. So all the rules for collecting data from a single population still apply. In addition, it is critical to store the information that identifies the population each subject unit comes from in a way that is as safe and error-free as possible.

In addition, for many purposes, in experimental studies, the information that identifies the population must be kept secret until data collection is complete. (This is called a *double-blind* study and ensures that the psychological predispositions of the people being studied, the people conducting the study, and the people collecting the data do not create bias and error in the results.) Population information may also be kept secret for privacy reasons. Often, the information that identifies the population from where the subject comes, together with their other data, is enough to reveal their identity. In these two sorts of cases, the population information must not only be stored safely, it must also be encoded and stored separately from the rest of the data and there must be a way to reconnect the two parts of the data when necessary.

In experimental studies, the methods used for the type of sampling may also be used to determine the population. For instance, in addition to random sampling, subjects may be randomly assigned to groups. (The group determines from which population the subject is assumed to come from.) In this case, every step in the random sampling and assignment procedures must be stored as well, in order to ensure an audit trail that reflects how each subject came from its respective population.

In quasi-experiments, there are a variety of methods used where a true control group, or a truly random assignment of members is not possible.

In some cases, there may still be multiple populations, and proper procedures for managing data and statistics from multiple populations apply.

Quiz

1. What is a disadvantage of using someone else's summary statistics?
   (a) We cannot double-check the calculations
   (b) We cannot calculate any new statistics
   (c) Both (a) and (b) are disadvantages
   (d) Neither (a) nor (b) are disadvantages

2. Free statistics...
   (a) Is always of the highest quality
   (b) Can be found everywhere
   (c) Is always worth what we pay for them
   (d) Always have the data available

3. _____ is the best insurance we can have of the quality of free data?
   (a) Data collection documentation
   (b) The assurance of the vendor
   (c) The source of any bias
   (d) Archived data

4. Non-archived data is data that is collected...
   (a) For the purposes of statistical studies
   (b) For reasons other than for statistical studies
   (c) By a vendor for the purposes of selling you the data
   (d) With the highest quality standards

5. What source(s) within our own company may be the source of data?
   (a) Financial records
   (b) Human Resource records
   (c) Computer data logs
   (d) All of the above

6. Dealing with missing data, erroneous records, and correcting recording errors in the data is...
   (a) Cleaning data
   (b) Coding data
   (c) Falsifying data
   (d) Analyzing data

7. The method for ensuring reliability in data entry is...
   (a) Data collection
   (b) Data redundancy
   (c) Data validation
   (d) Data calibration

8. The single most important aspect of database design is to...
   (a) Summarize the data as much as possible
   (b) Never throw any data away
   (c) Make good decisions about what data to record and what to throw away
   (d) Perform error correction

9. How does experimental and quasi-experimental data collection differ from survey data collection?
   (a) Data is sampled from more than one population
   (b) The rules of collecting survey data do not apply
   (c) Both (a) and (b)
   (d) There is no difference

10. A quasi-experiment occurs when...
   (a) A true control group is not possible
   (b) Random assignment is not possible
   (c) Either random assignment or a true control group is not possible
   (d) Both random assignment and a true control group are not possible

Statistics Without Numbers: Graphs and Charts

There are two critical uses for graphics in statistics: first, in order to perform statistical analyses, the distribution of the data *must* be examined visually, not only to ensure that the required assumptions for that particular statistical technique are met, but also in order to note any unusual characteristics of the data that might affect our interpretation of the results of the analyses. Second, graphics plays a pivotal role in communicating the results in support of the business decision, either in the written report or oral presentation.

Some types of graphs are not particularly useful for the first purpose of examining a distribution, but all types of graphs have their role in communicating results to an audience. This chapter shows you the sorts of graphs most useful for business presentations, indicates some of the uses of each, and gives guidelines for using graphs and charts to make a statistical report for business.

TIPS ON TERMS

Is it a Chart or is it a Graph? The terms "chart" and "graph" tend to be used more or less interchangeably, without a clear distinction between the two. In *Business Statistics Demystified*, *chart* and *graph* are two words for the same thing. We follow common usage, so we talk about pie charts and line graphs, rather than pie graphs and line charts.

When to Use Pictures: Clarity and Precision

Before we choose what type of graph or chart to use, we need to decide whether or not to use a chart or graph or other graphical element at all. There are really three ways we can tell our audience about quantitative (numerical) information. We can leave it as numbers (and put those numbers in a table), rephrase it as text, or convert it to a chart or graph.

Our goal is clarity, and graphics convey lots of information clearly. The decision as to how to present numerical information should be based, first and foremost, on how much information we need to present. If we can present all of the key information in a chart or graph, that should be our first choice. Even if there is additional information that is not as important as the key information, we should present the key information in a graph, and place the remainder of the less important information in a table, possibly stashing the table itself in an appendix.

KEY POINT

Graphs and Charts Support Business Decisions
Use graphics to support the business decision. Then, in your report, if you need to explain no more than six or seven numbers, put them in your text. If you need to show more numbers than that, use a table. And if you need to show all of your data, put the tables in the appendix. *Business Statistics Demystified*, itself, is a very good model of how to present information in figures, tables, and appendices.

Often, we think of graphics as improving clarity at the cost of a loss of precision. This is not really true. If our graph provides us with all the precision we need, then there is no effective loss of precision from changing from numbers to graphics. In addition, a well-designed graph can deliver a good deal more precision than we would imagine at the outset. Finally, even if the graph delivers less precision than we need, we can supplement the graph with a numerical table.

HANDY HINTS

A Picture and a Thousand Words
Graphs and words together make the most effective business presentation. And you don't need a thousand words. One to three well-written paragraphs per chart is about right. We've filled this chapter with a variety of examples of explanations of business points illustrated by graphs. As you read, learn to explain graphs, as well as to create them.

It is essential to use the right type of graph or chart. The wrong type of chart can easily make things much less clear and confuse the reader. In this chapter, you will learn which charts to use for different types of data and different presentation purposes.

CRITICAL CAUTION

Take Time to Learn to Create Good Graphs
Please read through this chapter to learn about the uses of different types of graphs. At the end, you will find a section called *Do's and Don'ts*. Read it carefully, or you may end up creating confusing or misleading graphs. Then grab a pen and paper and do a quick sketch of what you want your graph to look like. At that point, you will be ready to use—or learn to use—Microsoft Excel® or a statistical program that will generate graphs from numbers.

Parts is Parts: The Pie Chart

The pie chart is a dramatic way to show the proportions, that is, the ratios of several types that, together, constitute a whole. For example, the entire pie can represent our flock of sheep, with each pie slice representing one breed, as in Fig. 7-1. We could do another pie chart to show the proportions of groups of a different attribute, such as color. But it is essential that a pie chart divides

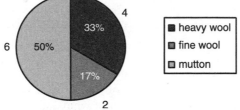

Fig. 7-1. Pie chart: breeds of sheep in our flock.

a whole into parts where each part is a different value of a single variable. So we would do one pie chart for breed, and another one for color. In business, a very common use of pie charts is for budgeting. The whole budget is the pie, and the slices can either represent different sources of revenue, or different allocations (but not both in the same chart).

HANDY HINTS

Requirements for a Pie Chart
- There must be an identifiable whole whose parts need to be identified
- Every single individual unit that is included in the whole must be uniquely identifiable as belonging to one and only one of the parts
- There must be a single, reliably measured variable that measures the proportion of the whole for each part
- The sum of the measures of the parts cannot exceed 100% of the whole
- If the sum of the largest and most important parts is less than 100% of the whole, the remainder of the parts must be grouped together to form a special category of "other." It must be possible to treat "other" category as a part, in a sensible and meaningful fashion. The "other" part is sometimes called "miscellaneous," and should be small. Some hold that it must be smaller than any other part.

As a general rule, the pie chart is best used for displaying counts of the different values of a single categorical variable as proportions of the whole. Because, in a pie chart, there is no set order to the values, it is used more often for nominal than ordinal variables. If the order of the values is important, another type of chart (perhaps a bar chart) should be used. If the comparison of values to each other is more important than comparison of each value to the whole, a bar chart should also be used.

HANDY HINTS

Serving up a Good Pie
When presenting a pie chart, be sure to define what the whole is. In statistics, be very clear whether the whole is the population or the sample. If presenting multiple pie charts, be sure to define what the whole is, and what the group of divisions are, for each chart.

Compare and Contrast: The Bar Chart

One of the most common types of chart is the bar chart. In a basic bar chart, there is one bar for each value of the variable being illustrated. The length (or height) of the bar indicates the count, called the *frequency* of each value of the variable. The bar chart allows us to look at the sample distribution of a single variable, but it also has other uses. There are many variations on the bar chart. Some have their own special names, others do not. Those that have been named may have more than one name, each derived from different fields. The bar chart can also be combined with other types of charts. In this section, we will take a look at some of the more common and more useful types of bar charts, and try to keep track of their names.

ORIENTATION OF THE BAR CHART

Bar charts can be either vertical, with the bars running up and down, or horizontal, with the bars running side to side. There are guidelines as to when one or the other type is best, but there are no hard and fast rules.

Vertical bar charts

The vertical bar chart is more common, for a number of reasons. In many cultures, the difference between up and down has much stronger connotations than does the difference between left and right. The length of a bar in a bar chart is the measure of the count or quantity of that particular value of the variable. When the bars stand vertically, bigger counts and more quantity make a taller bar. This fits nicely into our cultural notion that taller means bigger or more important. Take a look at Fig. 7-2 for an example of a vertical bar chart.

Note that this chart displays exactly the same data as our pie chart, Fig. 7-1. However, it displays it with a different focus. Here, instead of showing

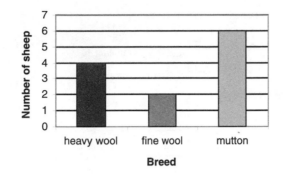

Fig. 7-2. Vertical bar chart: breeds of sheep in our flock.

proportions of the total, we are showing the counts in relation to one another, but not in relation to an image of the total. When displaying the same data in two or more different figures, it is useful to keep the color or shading consistent from one to the next, as we have done here.

Horizontal bar charts

When the length of the bar indicates something other than count or quantity, a horizontal bar chart should be considered. Time, for example, is best understood when laid out on a horizontal axis, often called a timeline. For example, the field of project management uses a variety of specialized horizontal bar charts, including the Gantt chart, to lay out project schedules over time.

Another important use of the horizontal bar chart is in the *side-by-side Bar Chart*. In the side-by-side chart, also called the mirrored chart, two distributions are compared by placing two bar charts, reflected back-to-back. If we use a vertical bar chart, then one of the two distributions points upward and the other points downward. This gives a strong impression that the first distribution is a measure of something positive or "good" and the second is a measure of something negative or "bad." Unless this is what we want to communicate, we should place the two distributions side-by-side with horizontal bars. Figure 7-3 uses a horizontal bar chart to show the percentage distribution of stocks in a personal portfolio compared to the distributions in the S&P 500. In general, people make this comparison to improve their investment strategy. Variance of one's personal portfolio from the S&P 500 might be a matter of investment history or personal preference. It could also be an indicator of conservatism (more manufacturing and retail), or a less risk-averse approach (more technology stocks). If you find that your portfolio has had significantly better—or worse—performance than the S&P 500 during the prior year, a table like this could help you

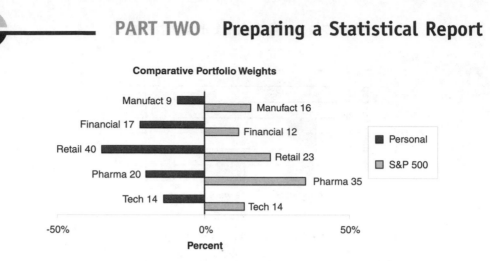

Fig. 7-3. Side-by-side bar chart: comparative portfolio weights.

identify reasons for the success—or issues—of your strategy. Just glancing at the chart, we can see that this portfolio has significantly less investment in pharmaceuticals, and somewhat less in manufacturing, than the S&P 500, while holding more in retail and financial, and matching the S&P 500 in technology stocks. (Note: the figures for the S&P 500 are illustrative, and do not actually represent that index.) On the graphical side, it would be better if the items lined up exactly with one another. Unfortunately, Microsoft Excel® does not offer a side-by-side bar chart option, and we dummied this up using a grouped horizontal bar chart with negative numbers.

Figure 7-4 illustrates why we do not generally use vertical mirrored bar charts. Here, we see a diagram showing the difference of ethnic diversity and opportunity between the free peoples of Middle-Earth and the dominion of the Dark Lord, Sauron, in J. R. R. Tolkien's *The Lord of the Rings*. In each society, those who achieve the highest honor are members of "the nine." For the good peoples, this is the nine walkers; for the evil Land of Mordor, this is the nine riders. If we note that there is, in fact, a diverse population of goblins

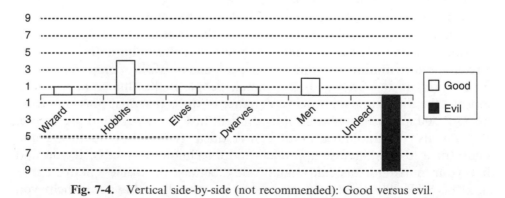

Fig. 7-4. Vertical side-by-side (not recommended): Good versus evil.

(or, to use the more proper term, orcs), trolls, evil men, evil wizards, spiders, and others who all support the Dark Lord, we realize that this is a society with a glass ceiling. Sauron will allow only undead to rise to the highest positions of power, leadership, and recognition, becoming members of the nine riders. While statistics cannot establish causality, it may be the case that this inequality of opportunity is one of the aspects of Sauron's realm which causes it legitimately to be considered evil.

PARETO CHARTS

The first specialized bar chart we need to look at is the Pareto chart, also called a Pareto diagram. A Pareto chart is just the name for a bar chart of values of a nominal variable. In the Pareto bar chart, the position of each bar along the x-axis (or, in the case of a horizontal chart, the y-axis) is arbitrary, or at least does not convey order. Figure 7-2, our vertical bar chart, is a Pareto chart. Since there is no order to the breeds, we could pick any order we wanted for the order of the bars in Fig. 7-2. In this case, we put the wool sheep next to each other. Very often, in a Pareto chart, the bars are ordered from tallest to shortest, or from most important to least important. You can find an example of a Pareto diagram of that type in Chapter 17 "Quality Management."

HISTOGRAMS, AREA CHARTS, AND FREQUENCY POLYGONS

A histogram, also called an area chart, is a type of bar chart used to show the frequency distribution of a numerical (interval or ratio) variable. (Some textbooks define a histogram differently and may specify somewhat different rules for constructing them.) There are two key rules for histograms (and a number of other less important standards).

First, the data must be *grouped* into equal-sized intervals. The data should be sorted in order using the values of the variable. Based on N, we need to estimate the number of intervals. So long as N is between about 50 and a few hundred, we should pick a number between 10 and 20. Next, we divide the range by the estimated number of intervals. If the result is not a whole number (integer), we choose the next largest whole number. This number is the interval size. Now, we choose the highest number, less than or equal to the minimum, that is evenly divisible by the interval size. That number is the bottom of the lowest equal-sized interval. The data can now be grouped. The length of each bar will be proportional to the number of units whose value for the variable is within each interval.

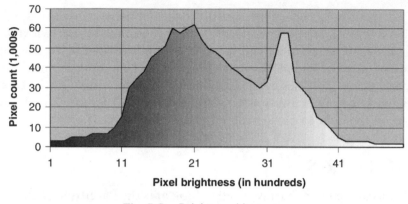

Fig. 7-5. Brightness histogram.

Second, the width of each bar is made equal to the interval size. Each bar is centered over the midpoint of the interval on the x-axis. This means that there will be no space between the bars.

When we eliminate the spaces between bars and use equal intervals, the bars have a total area equal to our total sample size, and we see the distribution of the sample across the X-axis. If we connect the tops of the bars with a line, instead of using separate bars, our histogram becomes a special kind of line chart called an *area chart* or *frequency polygon*. In fact, some people use the terms interchangeably. Figure 7-5 is called a brightness histogram because that is the standard term in the digital photography industry, but the actual figure is an area chart, because we used Microsoft Excel® to prepare charts for this book, and Excel® has area charts, but does not generate bar charts with no space between the bars.

Digital cameras produce a histogram that is an excellent illustration of their function. The X-axis measures brightness of pixels (picture elements), from solid black to pure white. The Y-axis is the count of pixels with that level of brightness. A good photograph has a majority of pixels in the middle of the range, not too dark, nor too light. Figure 7-5 provides an example of a histogram based on the pixel count from a photograph. A photograph's brightness histogram is one measure of the quality of the photograph. A sophisticated brightness histogram, like this one, shows the pixels on a gray scale from black to white. If most pixels are in the middle 60% of the image, and there is some interesting variation in that central area, the picture is likely to be pleasing. It will have a nice range of grays or middle colors, plus some interesting contrast. If an image is too dark or too light, we see it in the histogram, and we can use a computer graphics program to adjust the brightness of the image until it has a good histogram, and looks good, as well. Unfortunately, adjusting the graph of our earnings portfolio in a computer program doesn't have a similar effect on the value of our portfolio.

The purpose of the histogram, like the stem-and-leaf discussed in Chapter 3 "What Is Probability?" is to illustrate the shape of the sample distribution. Critical to this is that the data be grouped into the right number of intervals. As we saw with the stem-and-leaf, too many small intervals flatten the data out across the graph and hide the differences in counts across the range that are, in essence, the distribution. Likewise, too few big intervals stack the data up into the middle of the graph and also hide the differences. If the ups and downs of the distribution are not clear in the histogram, change the number of intervals and try again.

KEY POINT

Any graph designed to show a sample distribution needs to emphasize, visually, the changes in counts across the range. Key to creating this visual emphasis is the interval size used to group the data, which is determined by the number of intervals.

Histograms and frequency polygons have a function in the evaluation of statistical results. They are used to show—and to visually analyze—the distribution of the values of a single variable. For histograms with many intervals (that is, many bars), the line graph looks smoother and begins to look like a bell curve if the distribution is bell-shaped. Also, a frequency polygon can also be smoothed to approximate the shape of the population distribution from which the sample was drawn, in part because it retains the all important relationship between area under the line and the proportion of counts in each part of the range. Using these visual approaches, we can examine our data and see the mean and kurtosis, and see if we find unusual variation or truncation, as described in Chapter 3. If we see these effects, we can perform the appropriate statistical analyses to learn more about them.

SEGMENTED AND STACKED BAR CHARTS

A segmented bar chart is a way of displaying the relationship between two variables. We create a bar chart for the first variable. Then, we split each bar across, and use two different shadings within each bar. The shading subdivides the first variable according to how many of the individuals within that variable have each of the values of the second variable. Ordinarily, the second

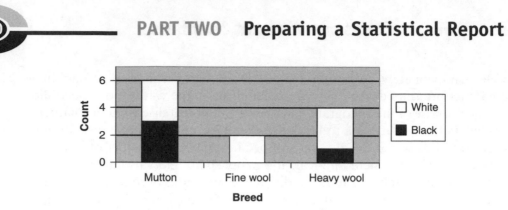

Fig. 7-6.   Segmented bar chart: sheep breeds and colors.

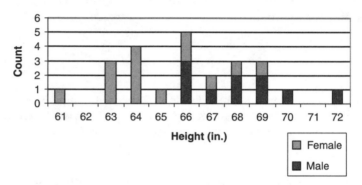

Fig. 7-7.   Stacked bar chart: male and female height distribution.

variable should be nominal and not ordinal, since shading is not a good way to show order. In Fig. 7-6, we have a segmented bar chart of our sheep flock, with the first variable being breed and the second variable being color (black or white). If we take a look at the third bar, it tells us that, of the four mutton sheep, three are white and one is black.

For a more sophisticated example of a segmented bar chart, take a look at Fig. 7-7. This shows the height of people, split out as men and women. This is a good example of the uses of a segmented bar chart. We see how many people are of each height on the chart, and also how many women, and how many men. Because men and women are sensible divisions of the group people (on the variable, gender), the segmented bar chart makes sense. Note that a vertical segmented bar chart is often called a *stacked bar chart*. If you want to contrast a stacked bar chart and a multiple bar chart of the same data, compare Fig. 7-7 with Fig. 3-4.

MULTIPLE BAR CHARTS

A multiple bar chart is a different way of showing the relationship between two variables. It is a better choice when neither variable is nominal. We

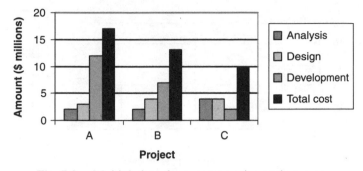

Fig. 7-8. Multiple bar chart: comparative project costs.

construct a multiple bar chart by dividing up our sample into sub-samples, based on the value of the *second* variable. Then we construct separate bar charts for each sub-sample. The multiple bar chart in Fig. 7-7 has the bars from each sub-sample interleaved across the x-axis.

Figure 7-8 could be used to illustrate the 1:10:100 rule, as follows. In Fig. 7-8, we see the expenses of three projects, broken out by project stage. The black bar, indicating total project cost goes down. Why? Because, as we moved from Project A to Project B to Project C, we put more effort into earlier planning. Project A represents our old "just do it" approach. In Project B, we spent more time and money on design, reducing rework during development, lowering development cost, and lowering the project cost. Encouraged by this, we tried out a new project approach with a great deal of effort for planning in both the analysis and design stages. The result was a much shorter development stage. These projects were of similar size and value to the company. Yet the improved methods used in Project C delivered the same result as project A for $10 million instead of $17 million, a savings of $7 million dollars, or 41%.

A multiple bar chart is also a good way to compare the distribution of a single variable for two or more samples that are parts of a survey or poll, as in Fig. 7-9. Instead of constructing sub-samples from a second variable, we just use the multiple samples. By looking at the bar graph of each sample, in this case an opinion poll of employees, stating how much they liked the company, broken out by interviewer, we have a chance to evaluate the data visually. Given that two of the three interviewers got results pretty close to a normal curve, but Interviewer C got a truncated curve with no zeros, and mostly fives and fours, we might ask if something biased Interviewer C's work. If this was a survey done using a quota sample, as described in Chapter 2 "What Is Statistics?," we might suspect that Interviewer C used a biased method for picking subjects. Or perhaps Interviewer C asked the question in a way that—intentionally or otherwise—encouraged people to

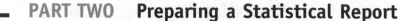
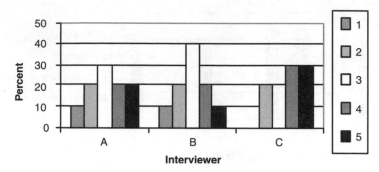

Fig. 7-9. Multiple bar chart to compare samples: survey results.

say they like the company. Or, perhaps, it wasn't Interviewer C's fault at all, and his work was timed when the annual bonus checks had just been distributed. In any case, there appears to be a bias, either in Interviewer C's results, or in the results of both Interviewer A and Interviewer B. A further—mathematical—statistical analysis would show us if the bias is significant enough that we can be assured that it has some cause, and is not within the range random variation.

Change: The Line Graph

The basic line graph shows the relative frequencies of values of a single variable, just like the bar chart. However, by connecting the counts for each value with a line, the reader's eye is drawn to the *change* in frequency from one value to the next. Because the focus is on the relationship between adjacent values of the variable, the line graph rarely makes much sense for nominal variables, and we should avoid using it for them. For ordinal, interval, and ratio variables, the line in a line graph shows the change in count across the order of the values of the variable.

SINGLE-LINE GRAPHS

The simplest way to think of a single-line graph is as a chart where a point is placed at the top of the center of each bar in a bar chart, all of the bars are erased, and the points are connected with a line. You can see an example of a single-line graph in Fig. 7-10. This shows number of customers served on Saturday nights over a 4 month period, January through April. The spike on February 14th is due to the popularity of fine dining on Valentine's Day. The line emphasizes the general trend of the growth of the restaurant's popularity.

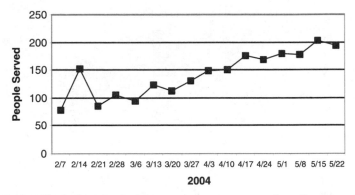

Fig. 7-10.   Single-line graph: restaurant customers served on Saturday night.

Line graphs can be modified in many ways. Two of the most basic changes are the choice of whether to show the points, or to show only the line. We can show the points if presenting specific figures is important, or to illustrate our data set. In Fig. 7-10, we used the points because the data are for each Saturday night, and not for the entire week. In our report or presentation, we would be sure to say that the line shows the trend for Saturday nights, not the number of people coming on other nights of the week. The other change is that, for some trends, it is appropriate to smooth the line into a curve, instead of a series of disjointed line segments. We do this when the shape of the curve is significant. We smoothen the curves for two reasons: to show a forecast; or to show whether our data are normally curved, biased, or truncated.

MULTI-LINE GRAPHS

A multi-line graph is just a combination of multiple-line graphs, one for each of a number of variables, superimposed on one pair of x and y-axes. It is very important that all of the variables being graphed have the same range of values. This is best for interval variables that only take on a small number of values, usually integer values. Figure 7-11 is a multiple-line graph that illustrates a growing business with a problem. Business, as measured by net revenue (in white), is growing nicely every quarter. Unfortunately, expenses (in gray) are growing to match, and as a result, net income (in black) is flat. Net income is net revenue less expenses. This graph illustrates how a graph can help focus a business discussion on the right issue. This graph makes it clear that net revenue is growing well, but that there is a problem because expenses are rising as fast as net revenue. This graph answers the question: Where do we have to look to find out why net income isn't growing?

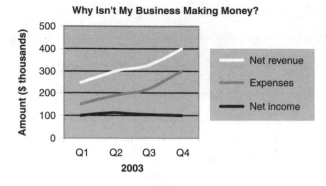

Fig. 7-11. Multiple-line graph: Why isn't my business making money?

The answer: we need to take a look and see why expenses are growing as rapidly as net revenue.

AREA GRAPHS

An area graph is a multi-line graph, where none of the lines cross and the difference between the frequency of each variable for each value is important. The differences in frequency are highlighted by adding different shadings between the adjacent lines of the graph. Note the confusing terminology: an area chart is a histogram, with only one variable, while an area graph is a line graph with multiple lines where the areas between the lines are important.

We could have used an area graph for our Fig. 7-11, Why Isn't My Business Making Money? We would have left out the black line (net income) and shaded the area between the two other lines black, to show that it's height did not vary. This would focus attention on the area with the unchanging height, net income, which is the problem we want our readers to focus on. For an example of an area graph, see Fig. 7-17.

Comparing Two Variables: The Scatter Plot

The scatter plot is a type of graph used to show the relationship between the distributions of two different numeric variables. (Scatter plots are most valuable when both variables have real number, rather than just integer, values.) Scatter plots are different than the other graphs we have looked at so far, because each and every unit is shown as a dot on the graph. For each individual subject unit, a point is placed on the graph. Each point is placed directly above the position on the x-axis equal to the unit's value on the first variable and directly to the right of the position on the y-axis equal to the unit's value on the second variable. The result is a graph with N points scattered over the chart.

What does a scatter plot show? A scatter plot shows how being high or low on one numeric variable relates to being high or low on a second numeric variable. This sort of relationship is very important to statistical techniques such as correlation and regression, which we will discuss in Chapter 12 "Correlation and Regression." If two variables measure features of the world that are involved in some of the same causal processes, then that can result in relationships that can show up in a scatter plot. To illustrate this, here are two scatter plots of Judy and her friends, Fig. 7-12, showing the relationship between their weights and their heights, and Fig. 7-13, showing the relationship between their weights and their I.Qs.

Note that the cloud of points in the first scatter plot is oval shaped and the cloud in the second plot is rounder. An oval cloud is an indication that there is a relationship between the two variables. As we might expect, taller folks

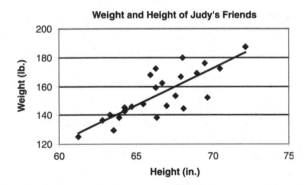

Fig. 7-12.   Scatter plot of weight and height of Judy's friends.

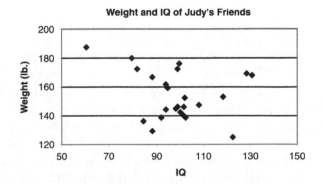

Fig. 7-13.   Scatter plot of weight and I.Q. of Judy's friends.

tend to weigh more. That means that, across a population, there are fewer folks who are tall and light (points on the upper left) or short and heavy (points on the lower right). (Even tall, skinny people weigh a fair amount because they are tall. And short, overweight people do not weigh as much as a tall overweight person might.) Because there are fewer points on the upper left and lower right, the overall cloud of points looks more oval. In short, the relationship between weight and height shows up in terms of the shape of the cloud of points in the scatter plot. We have gone one step further and created a line, called a regression line, on this scatter plot. Microsoft Excel automatically performed a linear regression for us and inserted the line. The regression line is a description of the relationship between the variables, and is discussed in Chapter 12.

In the case of weight and I.Q., where there is no relationship, we get a roughly circular cloud of points. This indicates that there is no visible correlation between the two variables; that is, they seem to be independent. If that happens, we do not draw a regression line. The set of points in Fig. 7-13 may not look very circular due to our small sample size. But it certainly does show that there is no indication of a correlation.

In some data sets, the regression line could slope downwards instead of upwards. Picture this chart: If we have two variables such as hours of television watched per week on the X-axis and grade-point-average on the Y-axis, we will see an oval cloud that starts high on the left (less TV, higher grades) and ends low on the right (more TV, lower grades). The regression line would slope from the upper left to the lower right, indicating a significant negative correlation between the two variables.

Don't Get Stuck in a Rut: Other Types of Figures

There are many more types of statistical charts than can be shown in this one chapter. There are box-and-whisker plots, pictograms, ideograms, digitdot plots, cross diagrams, bubble charts, contour plots, diamond charts, and statistical maps, among others. Some are specialized developments from one particular industry, or for one particular statistical application. It is good to have a nodding acquaintance with many different types of charts and graphs. It is important to be familiar with all of the types of charts and graphs used in our own industry. And, most important of all, we must understand how to use the basic charts well, and how to use the sophisticated charts of our industry making full use of all their features.

When the occasion arises, we should be able to scout around for the best type of graph to use. Often, the best chart doesn't fit into any of the standard categories, but may be a mixture of different standard chart types, or inspired by aspects of standard types we have seen. Here, we will look at mixing chart types, using 3-dimensional graphs, and creating statistical maps.

MIXING CHART TYPES

One of the most common ways to create a non-standard chart is by combining different chart types. Two different variables that extend over the same range may be compared by showing one with bars and the other with a line. The bubble chart combines aspects of a scatter plot with an ideogram, by having the size of each point increase with the values of a third variable. So long as things don't get too crowded or complex, there can be real value in combining chart types.

VOYAGE TO THE THIRD DIMENSION: 3-D CHARTS

In print, on a slide, or on the computer screen, there is no real third dimension. When we talk about 3-dimensional (3-D) graphs and charts, we mean graphs and charts where shading or perspective is used to give the appearance of a third dimension. There are two principle purposes of 3-D graphs, readability and additional information.

3-D charts allow us to compare counts of two different variables, with the count shown vertically, and the two variables shown on two axes at right angles to one another. This is useful when we want to highlight the comparison of the variables. Take a look at the 3-D chart in Fig. 7-14. This

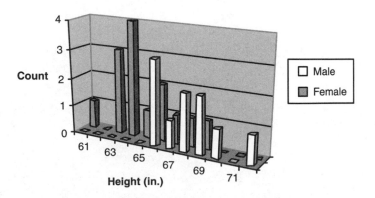

Fig. 7-14.   3-D Bar Chart: comparative frequency of heights of men and women.

chart makes a point visually: men are generally taller than women, but when we take into account variance (and not just the central tendency), many individual women are taller than many men. Compare Fig. 7-13 to the stacked bar chart Fig. 7-7. See how it is easier to make our point when women's and men's heights are put on the same 3-D chart beside one another.

Figure 7-14 has a vertical axis (or Z-axis) of count, an X-axis (heading to the right) of various heights of people, and a Y-axis (heading to the left and forward) of gender. A more advanced use of 3-D charts is to use the Y-axis to present a variable with more than two options. The 3-D variant on the line graph is called the *contour plot*. We can use a contour plot to see the interactive effects of two variables on a third variable.

In Fig. 7-15, the vertical axis shows the percent of product meeting customer requirements as a result of temperature variation and humidity variation during manufacturing. This chart might help with a business decision as follows. A company runs a manufacturing process in an old facility with poor humidity and temperature control. They are losing money due to a high percentage of rejected products. They want to choose among several options, from insulating the roof to acquiring a new building. These options have a wide range of costs, and they want to estimate the relative benefit of the different options over a ten-year period. Figure 7-15 shows the results of controlled experiments that shed light on the problem. Both temperature variation (in degrees) and humidity variation (in percent) during manufacturing have a significant deleterious effect on product quality. And the two interact in such a way that if both temperature and humidity vary, the cumulative effects are even worse.

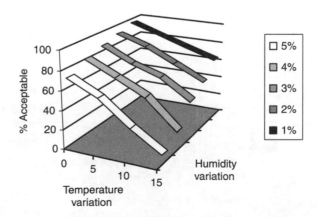

Fig. 7-15.   Contour plot: product quality by temperature and humidity variation.

This graph helps everyone at the meeting understand the problem clearly. Once everyone is on the same page, a discussion of the different solutions will be much easier than it would be otherwise. For example, an engineer might say, "just insulating the roof will not do a lot of good, because it doesn't fully control temperature variation, and it has no effect on humidity variation. If we insulate the roof and install air conditioning with humidity control, we can keep acceptability above 70%, because we will keep temperature variation within five degrees, and humidity variation within 2%. However, if we build a new facility, we can keep temperature variation to a fraction of a degree, and humidity variation to a fraction of a percent, maintaining a percentage of acceptable product above 90%." This technical information would feed into financial estimates. The financial estimates would compare the return on investment among the options leading to a recommendation for a solution.

CRITICAL CAUTION

Good 3-D Requires Care
There are three errors that are easy to make in preparing 3-D charts or graphs:
- *Hiding lower values behind higher ones.* Our readers don't have x-ray vision. Higher values on a contour plot or 3-D bar chart, when they are further to the left on the X-axis, or further to the right on the Y-axis, may hide the values behind them. Sometimes, we can manipulate the order of the axis to get around this, but we must then be very careful in explaining our chart. It might be better just to use two graphs.
- *Getting confused and confusing our audience.* To avoid this, think through variables and values.
- *Adding bias through 3-D.* Volumes give a different impression of variance or change than more representative 1-dimensional lines do.

MAPS

Among the specialized charts, the *statistical map* deserves special mention. When the values of a variable change from one place to another, we can show those changes on a map that shows both the places and the values. There are two common types of statistical maps. One simply uses shading to indicate different values or ranges of values as they apply to different places on the

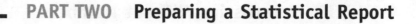

map. (For example, states with Republican governors might be in one color, and states with Democratic governors in a different color.) Another common statistical map shows 3-D bars like standing towers at the relevant locations. This could be used to indicate sales in different cities or regions.

Do's and Don'ts: Best Practices in Statistical Graphics

This section is a basic introduction to a good way to prepare graphics for business reports and to the most important errors to avoid. This section, alone, is not enough to allow you to be sure that you are not accidentally misrepresenting the data in your charts, especially if you choose to manipulate the axes, add 3-D effects, add color, or other flashy items. Read this section, read *How to Lie with Statistics*, be conservative in your style, and, if you can, have someone with more experience check your work. Otherwise, a chart can accidentally become an invalid, biased representation of the data all too easily.

DOING IT RIGHT

There are seven steps to including graphics in a statistical report with excellent results:

- *Plan the list of figures.* Here, we lay out the logic of our report.
- *Plan each chart or graph.* First, we prepare a chart type. Then we think through—and write down—the details.
- *Prepare each chart or graph.* Here, we generate the graph from the numbers.
- *Prepare text to go with each chart or graph.* Work with the examples in this chapter so that you can talk to your audience through each chart. Then write one to three paragraphs that do just that, whether people will be reading the report, or you will be giving a presentation.
- *Checking each chart or graph against the numbers.* Does everything look right? Check titles, axes, variables, and values.
- *Checking each figure against the table of graphs representing statistics.*
- *Do a final copy-edit and proofreading of the entire report, figures included.* Be sure to have someone else check your work. We catch each other's mistakes much better than we catch our own.

Several of these steps are worth a bit more explanation.

Planning the list of figures

If we make a logical list of the points we want to make, we can build a series of images that help our readers or listeners move from knowing nothing about a subject to knowing what we know. Each point should require a relatively small amount of information, and especially, it should not introduce many new types of information all at once. Once we have our logical presentation—or argument, or case, although we're not going to argue with anyone—we can look at each point and decide if it needs a graph, and what kind of graph would be best. If we skip this step, we are likely to find that we are creating very complicated graphs because we are packing too many points into one picture.

HANDY HINTS

Get Some Style and Save Some Time With Templates and Procedures

If we have several graphs of the same style—all line graphs or all bar charts, for instance—we should put them all in the same style, using consistent borders, fonts, line weights, and so forth. We can do this by creating a template, a procedure, or both. A template is a file with the basic settings for the graph all ready to go. To prepare each graph, we just drop in the data and make final adjustments. A procedure is a written series of steps that tells you—and others—what settings to choose in your graphics program to get the results you want.

The best approach is to make your first graph and play with it until it is exactly the way you want it. Then copy the axes and settings without the data to make the template, and note the steps and settings to write the procedure. Templates and procedures are particularly useful when graphics are being created by different people or with different tools. And, of course, you can keep them for your next project, producing high-quality graphs time after time.

To go to the last step, be willing to change style where it helps the presentation, but keep changes small. And, if you make an improvement that you like, be sure to change the template and procedure so you can use it the next time, and the next.

Planning each chart or graph

Here, we choose the type of chart or graph that we will use for each point we want to make in our report. Remember that we may not need a graphic at all, if plain text or a simple table will do. When we do need a graph, we can choose the type from Table 7-1.

Table 7.1   Chart types.

Chart name	Purpose or example
Pie chart	For showing proportional values of a variable in relation to the total.
Bar charts	
Horizontal bar chart	For comparing values in relation to one another, especially where we usually think of the item in horizontal terms, such as time, or where it is easiest to lay the graphic out horizontally.
Simple vertical bar chart	For comparing values in relation to one another, especially where we are comparing quantities or value.
Side-by-side or mirrored bar chart	For comparing two sets of values where the names of the values are the same, but the populations are different.
Segmented and stacked bar chart	The major variable is shown as a simple vertical bar chart. Each bar is then segmented into two or three parts, so that each variable is split into ranges, marked by color or shading, that indicate the values of another variable, most often a sub-category.
Multiple bar chart	Here, we cluster sets of bars in order, so that the chart can be read two ways to see how variables interact with one another. We might plot time, in quarters or years, and cluster divisions within each time or year, showing net income on the vertical axis. That would allow a comparison across time of each division, and a comparison across divisions within each time period.
Pareto diagram	A specialized bar chart for a single nominal variable.
Histogram or area chart	A chart—bar chart in the case of a histogram, or line graph in the case of an area chart—where the area shown on the chart shows the proportion of values in the sample. This requires meeting a number of specifications for the data and the graph. It allows visual inspection and representation of the sample for statistical and other purposes.
Line graphs	
Single line graph	To show how a single variable changes across the X-axis. Very often, the X-axis is time.

Table 7.1 Chart types (*Continued*).

Chart name	Purpose or example
Multiple line graph	To show how multiple variables change across the X-axis.
Area graph	To emphasize the areas on a line graph. We might want to emphasize one area that is, effectively, the subtraction of one value from another. Or we might want to compare several areas.
3-D graphs	
3-D Bar chart	To compare two different variables in relation to a third variable.
Contour plot	To compare the effects of two different variables on a third variable.
Other types of graphs	
Scatter plot	To evaluate the relationship between two different variables in a sample. This is specifically a statistical tool related to regression. Each sample is shown as a dot on an area, its location indicating its value for the variable of each of the two axes.
Statistical map	To show values of a variable across a region. This could be a geographical region. However, we can also consider some astronomical charts to be statistical maps of space, and it is possible to create statistical maps of other actual or theoretical regions, as well.
Stem-and-leaf	To get a quick look at the shape of a distribution, especially when no computer is available.
Box-and-whisker plot	To detect extreme values or other unusual characteristics of a sample distribution.
Bubble chart	To add a third variable to a scatter plot or map, where the third variable indicates a magnitude.
Contour plot	To show a third continuous variable that is dependent on the first two continuous variables.
Diamond chart	To show the relationship between actual performance on four variables against a standard.

CRITICAL CAUTION

Crucial Exceptions

There are times when a certain type of chart may seem like the right one, but using it would give the wrong impression. Be sure to read *Errors to avoid* when planning your figures.

Any list of options is incomplete, and prone to error. Think of Table 7-1 as a sample of the population of types of graphs you might use. There are some graphs you will encounter—especially the specialized graphs of your industry—which are not listed. In addition, there are some graphs that don't fall easily into our categories. For instance, we can put a line through a bar graph, to illustrate both value and change.

Also, the lengths of our bars and the heights of our lines do not need to indicate frequency. They can indicate count, or any numerical value. Line graphs can show cumulative values. There are many options, and the best way to learn them is to know the graphs in your industry, find good examples, and use them as templates. Another excellent way to learn is to find bad graphs, understand what is wrong with them, and fix them, as illustrated in our case study.

CASE STUDY

What Type of Graph to Use? The Case of the Bad Weather Graph

Picking the right sort of graph is a complex process. There are certain rules of thumb, but there is also a real need for applied common sense. Here is a lovely case of using a multiple bar chart when an area chart would be a much better choice. A prominent Cable TV channel that specializes in weather forecasts (we will mention no names!) also has a website that allows the user to display local weather and related information, including information about the local climate. The local climate, broken out by months, is displayed on the website in a multiple bar chart.

The original—in orange and blue—was a bit easier to read. Let that be a caution to those who rely on color in a world still full of black-and-white copiers and printers! First, note the poor spacing. When creating a multi-component bar chart, it is always a good idea to have all of the bars in each cluster close together—or even touching—and have greater separation between clusters. These allow the reader to see the clusters more easily. In Fig. 7-16, the only graphical element that tells you that two bars belong to the same month are the clumsy, distracting grid lines in the background.

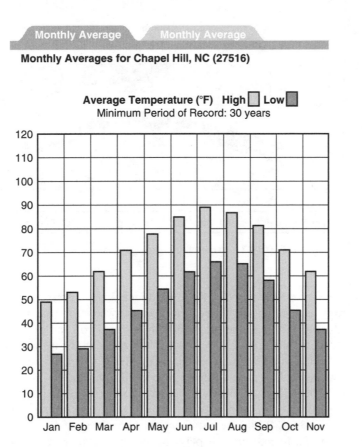

Monthly Averages for Chapel Hill, NC (27516)

Fig. 7-16.   Wrong choice of graph: monthly average temperature.

Second, bars are usually used to indicate amounts, and temperatures are not amounts of anything. Even absolute, or Kelvin, temperatures are not amounts of heat, and, on the Fahrenheit temperature scale, the zero point is arbitrary.

Third, bar charts, whether vertical or horizontal or histogram, are designed to compare relative values. We might want to compare average temperature between January and June (although that is unlikely), but we would never need to compare the average high temperature for a month to the average low. We know before we look at the graph. The high will be higher than the low.

Now that we've shown a bar graph is not a good chart, let's see why a line graph might do what we want. What is it that our reader might like to know from these data? Most likely, it is the trend of changing temperature across the seasons. Trends are better shown with line graphs. (Since trends in time are about change, always consider a line graph before a bar graph, when the x-axis is time.) Since we have two variables (high and low), we will need a multi-line graph. As we look at this graph, we see that the value of one variable is always higher than the other, so we should

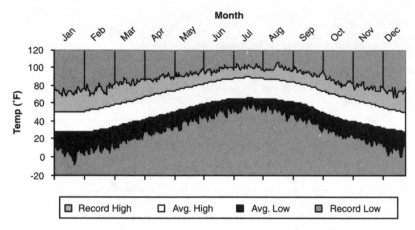

Fig. 7-17. Area chart: climate at a glance.

consider an area chart. An area chart is a good idea when the area between two lines on a line graph is meaningful. Here, the area between the average low and the average high for the month is the range of temperatures you can expect to feel on an average day. Above in Fig. 7-17, we have taken the data (kindly supplied by this same website) and created a simple area chart using Microsoft Excel®.

Note that we could have easily added the record lows and highs for each month, since they would be below and above the average lows and highs. We could probably have actually used daily, instead of monthly information, keeping only the months labeled along the x-axis. While the bar chart above looks crowded and cramped with only the little bit of information it displays so poorly, the area chart is so uncluttered that we could probably add more information without either distracting or confusing our readers.

There are tons of examples of bad graphs. This particular firm is doing no worse of a job than are many, many others. We should appreciate the bad examples, and use them as cautions and tools to learn to make good graphs.

ERRORS TO AVOID

There are so many ways to make errors in our graphics that the most important guidelines are to focus on making our point in a simple, clear way with the simplest, clearest graph we can use. Then, we use each part of the graph to represent the appropriate part of the information in our statistics.

Graphs representing statistics

Each element of a graph represents an element of the organization of our statistical data and statistical measures, as shown in Table 7-2.

Table 7.2   Graphical elements representing statistics.

Graphical element	Statistic represented	Do's and Don'ts
Title	The population or sample	Do • be clear whether you are speaking of a population or a sample Don't • leave out the title • use a cute, but unclear name
Primary axis	The frequency, count, or other value being focused on	Do • have the scale clearly marked with proper units. • for a ratio scale, have the axes cross at the zero point of the primary axis Don't • leave the zero point off the chart for a ratio scale • make a break in the chart between the zero point and representative values along the primary axis. (A marked break is sometimes acceptable, but introduces biased perception.)
Primary axis label	The variable shown on the primary axis	Do • include it • define units in parentheses, for example (in thousands)
Secondary axis	Shows the various values of the variable, or the most significant variable	Do • make it continuous or separated, whichever represents the actuality of the data • make a conscious choice of order for nominal scales • label the values, or show a scale, whichever is appropriate • place ordinal scales in order • show interval scales proportionally • show the zero-point of a ratio scale, either crossing the primary axis, or clearly labeled and perhaps marked with a vertical line • show appropriate intervals

Table 7.2 Graphical elements representing statistics (*Continued*).

Graphical element	Statistic represented	Do's and Don'ts
Secondary axis label	The variable shown on the secondary axis	Do • include it
Tertiary axis	Shows a variable less important than the one on the secondary axis. May be shown as a third dimension, or as the variable shown by segments or clusters	Do • on a 3-D graph, follow all the rules for a secondary axis • on another type of graph, use a key to provide all that information
Tertiary axis label	The variable shown on the tertiary axis	Do • include it
Key	Any information that cannot easily be shown by labels on the graph	Do • use a key whenever it makes interpretation easier, more certain, or clearer • proofread your key against the graph • use consistent keys and color choices across related graphs whenever possible

Clear labeling

There are a couple of things we can do to make our graphs very clear for our readers. One is to choose labels in the language of our reader—business language or everyday usage—and avoid statistical or engineering terms. We can create a glossary presentation terms to statistical terms in our appendix to show how we translated the statistics into usable English. Another is to make sure that our labels are a good size and in a good position in relation to the items that they identify.

Bias and nonlinear representation

Four manipulations of the Y-axis on graphs are quite common and should be avoided in most cases. They should be avoided because, even when clearly marked, they still introduce a visual bias that changes the reader's perception of significance. In general, do not:

- *Leave out the zero on a ratio scale.* This makes variance or change seem more significant than it is.
- *Have a break—marked or unmarked—on a ratio scale between the zero point and the values.* This also makes a variance or change seem more significant than it is. When all of the values are far from zero, but close to each other, it may be necessary to have a break in the scale. If so, always mark the break clearly. Better a break than to leave out the zero. If you do this, also note in the text that the heights of the bars do not indicate relative values.
- *Use nonlinear scales.* Logarithmic and other nonlinear scales have their place in engineering, but not in business. The visual meaning of items on these scales is not what it appears to the untrained audience, and they can be used to deceive, even causing lines that should be parallel or divergent to converge.
- *Use unfamiliar or undefined indices.* An index is a ratio. In business graphs, we can use familiar ratios that our audience uses all the time. We should avoid unfamiliar ratios. If we do include unfamiliar ratios, we must explain them carefully and define them consistently.

A more general form of statistical misrepresentation that often finds its way into graphs is the redefinition of the base of an index. We should avoid this scrupulously. Consider the following example. A company introduces a temporary 20% pay cut during difficult times. Your annual salary goes from $50,000 to $40,000. When things improve, the company gives everyone a 20% raise to restore their salaries. But 20% of what? A 20% raise on $40,000 is $8,000, and now you have your salary as $48,000, $2,000 below the original $50,000. Where did the $2,000 go? It was lost in the change of the base of the index. Twenty percent of $40,000 is $2,000 less than 20% of $50,000.

Cute pictures are often used to replace bar charts. The problem is that, when we do this, we create 2-dimensional or 3-dimensional images representing linear changes, which misrepresents them, making the change or variance appear much more significant than it really is. For good examples of this and other crucial cautions, see *How to Lie with Statistics*.

Too clever by half: the whens and wheres of color, cartoons, and photographs

As you become more sophisticated in preparing graphs, you may want to jazz them up a bit. After all, if your audience falls asleep during your presentation, you are not supporting a business decision very well!

However, we recommend that you focus on being relevant and clear first, and flashy a good deal later. If you come to the point that you want to make your charts fancier, we suggest you work cautiously. It is very easy to take creativity too far when showing data in a graph or chart. The colors that highlight the important features of our data rapidly become distractions when there are too many or too much. The clever combination of chart types that allow us to compare two variables may rapidly become completely confusing when we try to compare three. Use color and fancy types of charts to support the clarity of communications.

The same thing applies to appeals to the emotions. Charming cartoon characters can help make our point more memorable, but they can also distract. In addition, a poorly designed icon or ideogram can actually mislead. It is a classic, and misleading, error in graphing to have pictures whose size varies with some value when, actually, the height illustrates the difference in value, but the difference in area or volume gives a false impression of a larger difference than is actually there.

We will look into this further in Chapter 10 "Reporting the Results," where we discuss presenting statistical information in different business contexts, including decision support and advertising.

SOMETHING EXTRA

Technology: A Plus or a Minus?

Everything we have said about flashy effects, color, and emotional appeal applies even more to special effects we can add through computer graphics and new technology. Whenever a new technology becomes available, there is a tendency, called the "gee-whiz" effect, to overuse it. This is not a new thing; excessively gaudy print and photographs were used where a simple paragraph would have been clearer, all the way back in the 1800s. Snazzy changing color and moving pictures hide more than they reveal. They have no place in decision support, and their place in advertising based on statistics is questionable.

In addition, the more we rely on new technology, the more risk there is that things won't come out the way we planned. Could you present your report if the bulb blew on your data projector? What happens to your color-coded bar charts when extra people come to the meeting, and you are quickly making extra copies—on a black-and-white printer or copier? When we KISS—Keep It Simple, Sam—we also Play It Safe, Susie. (The acronym is left for the reader to develop.)

Quiz

1. Charts and graphs...
    (a)   May be useful for examining the distribution of the data
    (b)   Communicate the results of a statistical study
    (c)   Both (a) and (b)
    (d)   Neither (a) nor (b)

2. A pie chart is useful for illustrating _____.
    (a)   Proportions
    (b)   Comparisons
    (c)   Change in frequency
    (d)   All of the above

3. The length (or height) of a bar in a bar chart indicates the _____ of each value of the variable.
    (a)   Probability
    (b)   Ratio
    (c)   Proportion
    (d)   Frequency

4. A Pareto chart is used when the data are on a _____ scale.
    (a)   Ratio
    (b)   Nominal
    (c)   Ordinal
    (d)   Interval

5. We can use a _____ chart by dividing the sample into sub-samples.
    (a)   Line graph
    (b)   Histogram
    (c)   Multiple bar
    (d)   All of the above

6. The Line graph allows you to focus on the _____ from one value to the next.
    (a)   Distribution
    (b)   Frequency
    (c)   Ratio
    (d)   Change

7. A _____ shows each subject unit on the graph.
   - (a) Scatter plot
   - (b) Line graph
   - (c) Histogram
   - (d) Pie chart

8. Using templates or procedures when planning charts and graphs saves you...
   - (a) Time
   - (b) Cost
   - (c) Effort
   - (d) All of the above

9. The primary axis label on a graph should be...
   - (a) Included
   - (b) Defined in units
   - (c) Both (a) and (b)
   - (d) Neither (a) nor (b)

10. The use of color and flashy effects in graphs should only be used to...
    - (a) Add to the "gee-whiz" factor
    - (b) Provide clarity
    - (c) Distract from the results
    - (d) All of the above

8

Common Statistical Measures

We have reached the point where we are ready to talk specifically about statistical measures, one by one. Calculating statistical measures, like any other sort of calculation, means following a specific procedure. In the world of mathematics, procedures are specified using the special language of equations. We can't avoid equations in this chapter, but at least we can say why they are necessary. (If you are uncomfortable with equations or think you need a review of the basic rules, see Appendix A for a review of basic math.)

SURVIVAL STRATEGIES

The keys to getting the most out of equations are: First, remember that the symbol for the statistical measure being defined is always alone on the left of the equals sign. Second, everything on the right of the equals sign is just a short-hand for the rules for calculating the value of the statistical measure. The detailed version of those rules will be given in the adjoining text.

Recall that statistical measures are summaries. They are single numbers that describe some aspect or feature of a group of numbers. So, the procedures we will be using will be applied to many numbers and will end up producing one number. We will explain what feature of the group of numbers is described by each statistical measure as we go along.

Fundamental Measures

Most complex statistical measures are based on simpler ones. We calculate the simple statistics from the numerical data and then calculate the more complex statistics from the simpler ones. The two simplest and most basic statistics are the *count* and the *ratio*, which we saw earlier in Chapter 3 "What Is Probability?" These two statistics are so simple that most statistics books don't even talk about them. However, they are the basis for many, many other statistics, so we will discuss them here.

COUNTING: *N* AND DEGREES OF FREEDOM: *df*

The *count*, symbolized by *N*, is such a simple statistic that it doesn't even have an equation. We all learned to count before we knew any other sort of arithmetic. Counting is the most basic sort of procedure in mathematics and statistics. It is also the simplest sort of measurement. When we count something in the real world, we get a number. That sort of count is a measure, not a statistic.

Counting in statistics

But what happens when we count numbers instead of things? We start with a group of numbers and we end up with a single number that describes one feature of that group, namely, how big a group it is. When we count sheep, we get a number that is a measure of the flock. When we count numbers, like the number of the weights of all the sheep in the flock, we are calculating the statistic, *N*. (Of course, this is a distinction without a difference. The number of sheep will be the same as the number of numbers we get when we weigh them.)

The count of all the numbers in a group of numbers is the basis for almost every other statistic. When we look at the equations for almost every other statistic, we will see N on the right-hand side of the equals sign.

Usually, we use the symbol, N, to indicate the size of the entire sample. When we are talking about smaller groups that are part of the sample, we usually add a *subscript* to the N to show the difference. For example, if we wanted to talk about only our black sheep, we might use $N = 12$ as the equation for the entire flock and $N_b = 5$ as the equation for the count of just the black sheep.

HANDY HINTS

Note that, in these first equations, we are not specifying a procedure. When there is just a number to the right of the equals sign, the equation is being used to specify the value of the variable in that particular case. Mathematicians (and statisticians) use equations for a lot of different sorts of things, and don't always warn us when they shift gears.

CRITICAL CAUTION

In statistical *theory*, when the population is finite, N is sometimes used to mean the size of the entire population. The symbol, n, is then used to describe the size of the sample. We won't be doing much theory here in *Business Statistics Demystified*, but elsewhere, this can get a bit confusing.

What are degrees of freedom?

There is a very sophisticated notion in statistical theory that we will need later on when we talk about more complicated types of statistical techniques. This notion is called the *degrees of freedom*. Even though it is a difficult idea, it is calculated from a very simple statistic, the sample size, n. Equation 8-1 shows that the degrees of freedom (df) is equal to the sample size.

$$df = n \tag{8-1}$$

Every time we observe the world, we gather information about it. In statistics, each observation contributes one unit to the sample. The sample size, n, is the number of units in the sample, and thus is a measure of how much information we have obtained about the world with that sample.

When we use *n* as a measure of the amount of information in our sample, we call it the *degrees of freedom.*

Now, suppose we calculate a statistic from our sample. That gives us one piece of information about the world, taken from the sample. As it turns out, there is an important sense in which each number we calculate is worth the same as each number we collect. So, when we calculate a statistic from the sample, we have one less piece of information in the sample.

At first, this may seem odd. After all, we still have all *N* numbers from our sample. We still know what they are. What has been lost? The answer is that we may very well want to calculate more statistics from the same sample. How many times can we use the same *N* numbers, our data, to calculate statistics and still be finding out about the world, instead of just spinning our wheels? The answer is that we can calculate *N* statistics from *N* numbers before we run out of information. As we examine various statistical measures throughout this chapter, we will take a few more looks at degrees of freedom. Degrees of freedom will also be very important when we get to Part Three and talk about statistical techniques.

RATIOS, ALSO KNOWN AS PROPORTIONS

We took our first look at ratios back in Chapter 3 "What Is Probability?," when we learned about samples. Ratios are another very basic statistic. Ratios are calculated using two counts and nothing else. Just as we said at the outset, each more complex statistic will tend to use the simpler statistics as a starting point. As we noted in Chapter 3, the symbol for a ratio is *p*, because a ratio is our best estimate of the probability that a single subject unit sampled from the population will have that characteristic value on that variable.

TIPS ON TERMS

In Chapter 3 we distinguished between the terms, *ratio*, *proportion*, *probability*, and *odds*. While other texts may treat this differently or not treat it at all, in *Business Statistics Demystified*, we use the term *ratio* to mean the underlying relationship between two values. *Odds*, which are not used as often in statistics, involve a different sort of calculation than do both *proportion* and *probability*. The difference between a proportion and a probability is that proportions are measures taken on samples that show a ratio, while probabilities are theoretical values related to events with populations. A proportion of a sample gives the best *estimate* of the probability, but the probability itself, because it is a theoretical aspect of the population, is something we can never know for certain. We will hear a good deal more about estimates as we go forward.

Calculating the ratio

While there are other uses for ratios, the ratio statistic is most often used to estimate probabilities by comparing the count of some subgroup of interest to the count of a larger group to which that subgroup belongs. (It does not matter here whether we think of these counts as counts of things or counts of numbers.) This is not as complicated as it sounds. When we have a sample, that is a group. If we have a categorical variable, any value of that variable forms a subgroup. Our example of this is our flock of sheep (a group) and the variable, color. The value, black, of the variable, color, defines a subgroup. The ratio of the count of black sheep to the total count of sheep gives us the ratio or proportion of black sheep. In general, if the value is represented by x, the equation defining the proportion is:

$$p = \frac{N_x}{N} \tag{8-2}$$

All we have to do in Equation 8-2 is divide the count of the individual units having a value of x by the count of all the units in the group. And that is our second statistical equation. See how easy that was!

It is important to note here that we can calculate proportions for other types of variables as well as categorical ones. Remember that numerical variables (either interval or ratio) contain *more* information than do categorical variables. It is a common trick in statistics to redefine a numerical variable as a categorical variable in order to use it for categorical things, including calculating proportions.

The trick to creating categories from a numerical variable uses ranges of numbers. While we have not yet discussed the range as a statistic, we already know about ranges from algebra. (Refer to Appendix A for a refresher on the number line if you need to.) Ranges of numbers are things like, *less than or equal to ten, greater than −8.34,* or *greater than or equal to 8 and less than 11.* For any numerical variable, we can define a category as when the value of a numerical variable falls within a particular range. (The complementary category is when the value falls outside that range.) We saw this trick used in Chapter 3 where the probability of a value was defined using the normal curve.

If we redefine x to be a range for a numerical variable instead of a value for a categorical variable, we get the definition of the ratio for values of numerical variables, without even having to use a new equation!

FUN FACTS

In probability theory, mathematicians use the concept of a *set* to define *x* so that they only have to use one equation for both types of variables. Even the mathematicians like to keep some equations to a minimum.

Estimating population values

Finally, we need to understand the critical importance of the ratio to the statistical process called *estimation*. By estimation, statisticians mean taking a statistical measure of a sample and inferring the value of that same measure for the entire population. Very often, the value of a statistic for the population is a terrific help in making a business decision. For example, knowing how many people in the general public might be interested in buying our product would be very helpful in our marketing plan. But we can't very well ask every person in the country what they think. We can, however, ask a *sample* of the population of the country what they think. If we can use those data to estimate how many folks in the whole population feel the same way, we will be better able to make our marketing decisions. As we move through this chapter, we will see how each of the various statistics can be valuable in making business decisions, but usually only if we can use them to speak about their value for more than just the sample we have taken.

Recall from High School algebra that ratios always mean the same thing, no matter what the size of what is measured. If we have the same amount of plain chocolate bars as almond chocolate bars, then the total number of almond chocolate bars depends on how many chocolate bars we have in total. If we have six chocolate bars, then three of them are almond. If we have 50,000 chocolate bars, then 25,000 of them are almond. It is the ratio called a "half" that describes what is the same about these two batches of chocolate bars that differ so much in size. No matter how many or how few chocolate bars we have, a half is always a half.

HANDY HINTS

One way to think about what ratios do in statistics is that they, in effect, *erase* the information about the size of the sample, by dividing the count of objects of interest by *N*, the sample size.

The importance of this feature of ratios to estimation is that, in the end, the size of our sample is unimportant in terms of what we want to know about the whole population. (Make no mistake, the size of the sample is vitally important to calculating the statistics we need to help make our business decisions, but it is not a piece of information about the world, only about our study, and thus is only important as a part of the process.) If we talk to a random sample of the population and find that one-third of the people we talk to like our product, then our best estimate is that one-third of the general population will like our product as well. Just as in arithmetic, one-third is one-third is one-third.

The ratio is the key to making samples make sense. The value of the ratio always falls between zero and one. Its value doesn't change the size of the total group, whether the total group is the sample we happened to take, or the group is the total population we want to know about. If we can express an important value as a proportion of the sample using ratios, statistical theory tells us that that same value will apply—more or less, given error due to sampling—to the entire population, and can even tell us how much error to expect. That is how the process of estimation works.

CRITICAL CAUTION

Sample Size Does Matter
When we make an inference from a ratio calculated on a sample to a proportion of the population, how do we know our inference is valid? Here, sample size is a key factor. The larger the sample size, the smaller the possible error of our estimate. If our sample size is too small, then we can't make any reliable estimates about the whole population.

Descriptive Statistics: Characterizing Distributions

In statistics, our sample is our only source of information about the world. Statistics we calculate from the sample will be used to estimate population values. Recall from our discussion of samples in Chapter 3 "What Is Probability?" that, for each variable, there are N values that make up our *sample distribution* of that variable. The most basic statistics that provide values that are estimates of population values are called descriptive statistics.

Descriptive statistics describe features of the distribution. Each sample descriptive statistic describes a feature of the sample distribution. If the sample is large enough, the distribution of the sample will resemble the population distribution. Because of this, descriptions of the sample distribution will also describe (via estimation) the population distribution. Descriptions of the population can be the facts we need to help with our business decisions.

HANDY HINTS

While some descriptive statistics can be defined for ordinal variables, the theory of estimation for descriptive statistics assumes that the variables are numeric. For the remainder of this chapter, we will only be dealing with numeric variables. (Estimation of categorical variables usually involves only proportions, as estimated by ratios, which we learned about in the previous section.)

THE RANGE: MINIMUM AND MAXIMUM VALUES

For ordinal and numerical variables, where every value is higher or lower than every other value, there is at least one value in the sample distribution that is the lowest and one that is the highest. The lowest value is called the minimum. The highest value is called the maximum. Together, these two values, called the limits of the distribution, tell us the *range* of values of the sample distribution.

CRITICAL CAUTION

In statistics, the term, *range*, is used to mean two different things. Mathematically, the range is all the numbers on the number line between the minimum and the maximum. The range is described with both numbers. The *size* of the range is the maximum minus the minimum. The size of the range is a single number, a statistical measure, which is also (confusingly) called the range. In order to be consistent with other statistics books, we will use the term, range, for both the pair of limits and for the algebraic difference between the two. When things get too confusing, we will refer to the range statistic (the single number) as the "size of the range."

In some ways, the range of a numerical variable tells us the most basic fact about the distribution. It tells us what values are and are not to be found in

Table 8-1 Heights of Judy and female friends (sorted).

Name	Height (inches)
Tammy	61
Caroline	63
Hannah	$63\frac{1}{2}$
Ng	$63\frac{1}{2}$
Judy	64
Angela	64
Rita	64
Wai-Jung	$64\frac{1}{2}$
Shizuko	$65\frac{1}{2}$
Betty	$66\frac{1}{2}$
Maria	$66\frac{1}{2}$
Donna	67
Liora	68
Francie	$69\frac{1}{2}$

our sample. For example, the range of heights of Judy's female friends in Table 8-1 is from Tammy at $5'1''$ to Francie at $5'9\frac{1}{2}''$. Knowing the range means knowing that none of Judy's female friends is $5'10''$ or taller.

The size of the range tells us something different. It is a statistic that tells us how spread out the values are. The size of the range, a.k.a. the range statistic, of the heights of Judy's female friends is $8\frac{1}{2}$ inches. The range statistic tells us how much the heights of Judy's female friends vary. However, it is not a very good measure of variability. Imagine that all of Judy's female friends ranged in height from $5'1''$ to $5'3''$, except for Francie at $5'9\frac{1}{2}''$. This would mean that, overall, the heights clustered together a lot more than they actually do, but the value of the range statistic would be the same. The range statistic is a poor measure of the variability of a sample because it only depends on two values. As a result, it doesn't change with different intermediate values—and doesn't tell us anything about them, either. Different patterns of clustering between the minimum and the maximum don't change the range.

The range is simple to calculate, as we see in Equation 8-3, and it is important because it relates to the distribution limits. It is not used much in statistical inference, in part because it is not a very good measure of variability, but also because the mathematics of estimating using limits is very hard.

$$\text{range} = \text{max} - \text{min} \tag{8-3}$$

The range gives us a good first example of how degrees of freedom work. If our sample size is $N = 2$, and we know the maximum and the minimum, there is nothing more to know. If we lost our original data, but knew the sample size was two, and also knew the minimum and the maximum, we could reconstruct our data from that information. On the other hand, if our sample size is three or more, we wouldn't be able to reconstruct all of the original data. The two statistics, the minimum and the maximum, contain as much information as a sample of size, $N = 2$. A larger sample contains more information, which is why N is a good measure of the amount of information, or degrees of freedom, of our sample.

As an example of using ranges in business, consider a craftsperson choosing what items to make to sell at two upcoming crafts fairs. She looks at her sales results from the same fairs last year, when she brought all of her items, at different prices, to two fairs. One, a Christmas fair, had a lot of college students buying gifts. The least expensive item she sold had a price of $1, and the most expensive had a price of $79. At a fair sponsored by the art museum, the least expensive item she sold was $10, and the most expensive was $350. As she plans what to do this year, she can decide to bring items in a price range of $1 to $100 for the college Christmas fair, and $10 to $500 to the art fair. (She figures she can bring a few higher-end items and cross her fingers that they will sell.) But if she knew how many items she sold at each price, she could plan her production even more accurately.

TYPICALITY: MEANS, MEDIANS, AND MODES

The range uses two numbers to tell us about where on the number line our values fall. The size of the range (one number) doesn't include that information. It only tells us about variability. Neither one is a terrific measure, because values in the middle can shift around and not change the range. Suppose we were interested in selling sports equipment to Judy and her friends. We might want to know if, overall or *on the average*, Judy's female friends are "tall," because that might indicate how interested they are in basketball.

Of course, "tall" is a very imprecise term. How tall is tall? In statistics, when we want to have a precise numerical measure of where the values fall, we speak of an *average*. An average is a single number that summarizes how big or small all the values of a variable are. Because an average, unlike the range, is a single number, we can anticipate that that number will be somewhere in the middle of our distribution. For the most part, that is true.

There are many, many types of averages. Three of the most common are the *mean*, the *median*, and the *mode*. An average value is a typical value for the distribution, with "typical" defined differently for each type of average. In statistics, measures of typicality are often called measures of *central tendency* because, when the distribution is bell-shaped, the most typical values tend to be in the center of the curve.

The mode

The mode, like the minimum and the maximum, is a specific value that is actually found in our data. The way that this value is typical is that it is the value that appears the most times in our sample distribution. The mode of the distribution of the heights of Judy's female friends is 64, because three of these women are 5′4″ and no more than two of the rest are the same height as each other.

There are four important things to know about the mode. First, the mode tends to fall in the middle of the range for bell-shaped curves, but less so, the less bell-shaped the distribution is.

Second, the mode is not a terrific measure of how big the values are because it only depends on a few values. As with the range, we could change a lot of values and still end up with the same mode, or we could change just a few values and get a completely different mode, with almost the same distribution. Suppose that Wai-Jun was an inch taller and Donna was just half an inch shorter. With just those two changes, the mode changes from 64 to $66\frac{1}{2}$, which is a big change in describing what is typical.

Third, when we calculate the mode, we don't always get a single number. Suppose that Wai-Jun was half an inch shorter. Then there would be three women who are $5'3\frac{1}{2}''$ tall and also three women at 5′4″. Both would be modes. This sort of distribution is called *multi-modal*, because it has more than one mode.

Fourth, the mode depends critically on how we measure our data. Suppose we had measured each of the heights to the 1/100th of an inch. There would probably be no two women with exactly the same height and there would be no unique mode at all. Every value would be a mode. Ordinarily, when we need to calculate the mode, we need to be careful as to how we group our data. As we discussed in Chapter 3 "What Is Probability?" on sample distributions, with numerical data, grouping is done by rounding. If the grouping is right for the graph of our distribution, it will probably give us a good mode. We can try out a variety of stem-and-leaf diagrams to choose the best grouping for our data.

The median

The median is usually, but not always, a specific value that is actually found in our data. The way that this value is typical is that it is the value found in the middle of the sample distribution, sorted by size. Of course, when we graph our distribution with a stem-and-leaf (or, as we saw in Chapter 7 "Graphs and Charts," a histogram or frequency polygon), we are effectively sorting the numbers in order by size. That is what makes the shape of each distribution what it is.

If N is odd, calculating the median is trivial. If there are three numbers, and we sort them (for example, in a spreadsheet), the second number is the middle number and that value is the median. If there are seven numbers, the fourth number in order is the median. For any odd N, the median is just the $((N+1)/2)$th value when the numbers are sorted in order. Just add one to N and divide by two. Count down the sorted list to that number and that's the median. For example, let's look at all of Judy's female friends, *not including Judy herself*. What is the median height for this group? Judy has 13 friends, so we want the $((13+1)/2)$th or 7th height in order sorted by size. The 7th height is $64\frac{1}{2}$, which is the median of that distribution.

If N is even, then things get a tiny bit more complicated. If there are four numbers, then the second number and the third number have equal claim on being the "middle" number. Similarly, if there are ten numbers, then the fifth and the sixth numbers are both in the middle. If both numbers are the same, then there is no problem. If they are different, we use the number halfway between the two. For example, if we consider Judy *and* her female friends, there are 14 numbers. The 7th number is 64, and the 8th is $64\frac{1}{2}$. The number halfway between them is $64\frac{1}{4}$, and that is the median.

The median, like the mode, is also affected by the grouping of data. There is a more complex formula for the median, which some textbooks teach. That formula is based on the notion that the true values are never identical, but are grouped together, either by imprecision in our measurements or by rounding. The heights of Judy's friends are a good example. We record heights as real numbers, but we usually measure them only with enough precision to record them to the half inch or so. (A person's height changes as much as a half inch throughout the course of the day, and probably changes a few millimeters just from breathing, so more precise measurements of human height don't really make much sense.) We can, therefore, feel confident that Judy, Angela, and Rita are not all exactly the same height, even though the data have them all at 64 inches. The formula for the median with grouped data assumes that the real heights of the three women are evenly spaced from $63\frac{3}{4}$ inches to $64\frac{1}{4}$ inches, which is the range over which we would round up or down to 64. Of course, that assumption may not be true, but the formula at least takes into account what we know to be true, that no measurement (other than counting) is perfectly precise. To keep things simple, we will not be using this more complex calculation of the median in *Business Statistics Demystified*.

The mean

Unlike the mode and the median, the mean is almost never an actual value found in the original data. (We will see why in a minute.) The mean is the most commonly used type of average and probably the best known formula in statistics. It is also the first full-fledged statistic that has all of the numerical and statistical properties we will see in most of the more complex statistics from here on out. Three of these properties are worth looking at now. The mean is *sufficient*. It is a *parameter*. And it is a *moment*.

A sufficient statistic is a statistic that uses all of the information available in the sample to calculate the feature that statistic describes. The mean, like the mode and the median, describes what a typical value looks like. Calculations for the mode and the median attempt to locate a single value in the data that is typical. This is why other parts of the data can change without affecting the value of the mode or the median. Calculations for the mean involve using all of the data to produce just one representative number. This is why the mean is rarely an actual number found in the data. It is *like* all of the numbers, but usually isn't exactly the same as any of them.

The trick for seeing if a statistic is sufficient is to see if there is any way to change the value of just one piece of data, even by a little bit, and still get the same value for the statistic. As we have seen for the mode and the median, this is true for them. We can change just one value, and the mode and median might not change. So they are not sufficient statistics. As we will see soon, if we change just one value—even by a tiny bit—we will change the mean. And that is what makes it a sufficient statistic.

CRITICAL CAUTION

There are ways to change the data and not change the value of a sufficient statistic. For instance, for the mean, if we raise the value of one number a bit and lower the value of some other number the same amount, the mean will stay the same. The rule for sufficiency is that you cannot change just *one* number even a little bit without changing the value of the statistic. Think of it like the old slogan for Lays® potato chips: "Betcha can't change just one!"

The mean is also a parameter. The first thing to know about a parameter is that it can be calculated both for the sample and for the population, although usually not using the same formula. Remember that N was a characteristic of the sample, and had no corresponding value in the population (unless the population is finite, which we can't assume). The maximum, minimum, and range also do not necessarily have population values, even though we can always calculate them for the sample. The mode can always be calculated for the population, but the relationship between the mode of the sample and the mode of the population is highly unpredictable and not very useful. The sample median, as we calculated it above, does not have a corresponding formula for the population. The sample median for grouped data, which is the more complicated formula given in some textbooks, *does* have a corresponding formula for the population, but it is mostly used in specific kinds of statistical procedures, like some of the ones we will discuss in Chapter 14 "Nonparametric Statistics."

The mean, on the other hand, not only has a definition and a formula for the population, but there is a specific relationship between the way we calculate the sample mean and the population mean. As a result, mathematicians have proven that the sample mean will be a good estimate of the population mean, which is critical for statistical inference.

The other thing that makes a statistic a parameter is that the theoretical formula for some types of population distributions uses that statistic to define differences between distribution curves of the same type. For example, if the population distribution is normal, that means that not only is it bell-shaped, but it is a very specific bell shape. However, as we saw in Chapter 3 "What Is Probability?" a normal curve can lie on the left or on the right on the number line, and it can be skinny or fat and still be *normal*. The way that we distinguish between two different normal curves is by using numbers called *parameters*.

As it turns out, the *only* two ways a normal curve can be different and still be normal is by moving to the left or right, and by getting fatter or skinnier.

Where on the number line the normal curve lies is defined by the horizontal position of the middle of the curve, which is also the highest point. As it turns out, the middle of a normal curve also happens to be (you guessed it!) the value of the population mean. So, the population mean is the first parameter of the normal distribution. The second parameter is the variance, which we will be discussing in the next section.

As we mentioned earlier, there are many, many different kinds of theoretical distributions besides the normal distribution, but they all have a few things in common. They are all defined over the number line *using a formula*. And that formula contains *parameters*, which allow us to define a specific example of that type of distribution that has a slightly different shape or location or whatever. Parameters are very important to statistical inference. The general idea is that, if we know what *type* of distribution the population is, then all we need to know is the values of the parameters to know everything we can know about the population. If, as in the case of the mean, we can estimate the population value (the parameter) using the sample value, then we have a direct way of using our data to calculate valuable information about the rest of the world.

Finally, we said that the mean is a moment. Moments are less important in statistical inference, but they are helpful in understanding the important notion of degrees of freedom discussed earlier in this chapter. Geometrically, a distribution is a curve drawn above the number line that shows the relative likelihoods for values at each point. There is an infinite series of statistics, called moments, which can be used to specify precisely any distribution, even if the distribution cannot be defined with a theoretical formula. Each moment captures a bit of information about the overall shape of the distribution not captured by any of the previous ones. Together, all of the moments describe the shape of the distribution entirely.

In principle, we could calculate all of the corresponding sample statistics for each moment and determine the exact shape of the population distribution. (This is actually a little-used statistical inference technique called "the method of moments.") The problem is that there is no way to calculate an infinite number of statistics from a finite amount of data. The degrees of freedom of our data, equal to the number of values, N, sets a maximum on the number of sample statistics we can calculate without exhausting the information in our sample.

Because our data are so often bell-shaped, or else fall into some other well-known and well-defined shape, it turns out to be a better idea to assume that we know the formula for the distribution and just estimate the two or three parameters, rather than an infinite number of moments.

Now that we know what the mean does, and why it is so important, we can learn the exact formula. Equation 8-4 gives the formula for the sample mean:

$$m = \frac{\Sigma x}{N} \tag{8-4}$$

This equation requires a bit of explanation. The symbol, m, stands for the sample mean. We will see the equation for the population mean below. After the equals sign, we see a new symbol, Σ, which is the upper-case Greek letter, *sigma*. Sigma is the Greek equivalent for "S" and, here, in upper-case, it stands for sum. Σ, called the summation sign, is a standard shorthand notation for adding a whole bunch of numbers. It is not only easier to write than a long series of plus signs, but it also has the additional advantage that it tells us to add a bunch of numbers, even if we don't know exactly how many numbers we have to add.

In *Business Statistics Demystified*, when we use the summation sign without anything above or below it, we mean that the values of the part of the equation after the sign should be added for every individual unit in the sample. In this case, following the summation sign, we have x, which stands for the variable. So, Σx means we should add up the values of the variable for every unit in the sample. (There is also an extended summation sign, which allows us to specify exactly what values to add up.) In the case of Judy and her friends, this means adding up all the heights to get the sum total.

Below the division sign, in the denominator, we have our old friend, N, so, the procedure for calculating the sample mean is simply to add up all the numbers and divide by the number of numbers (the count). This is what we ordinarily call the "average" when we aren't doing statistics. The mean of the height of Judy and her female friends is 65 inches or 5'5″, which is not the height of any of those individual women.

The benefit of having the exact equation is that we can understand the statistic better. For example, we can see that the mean is sufficient. If we change any one value even a little bit, that will change the sum, but it won't change the count. That means it will change the mean by a small fraction of a bit, so the mean is sufficient. We can also see why the mean is not usually one of the original values from our sample. We have added all those numbers up together and there is no easy way to undo that.

HANDY HINTS

If we think about it a bit, we can use the equation to see why the mean gives us a measure of typicality. When we add all the numbers together, we mix the little numbers in with the big ones. The little ones make up less than their fair share of the sum and the big ones make up more than their fair share. And what is their fair share? There are N numbers, so a fair share would be one Nth of the sum, which is the mean, exactly. This is easier to think about with values we can exchange, like money, rather than things like heights. There is a joke that runs like this: Three men are sitting in a bar. Bill Gates walks into the bar. The median salary of customers at the bar jumps to over a hundred billion of dollars, but that doesn't help the three guys who got there first. On the other hand, if everyone in the bar agrees to share their money equally, putting it all into one pot, and giving each patron $\frac{1}{4}$ of the pot, the three men would be much better off.

KEY POINT

There is also an equation for the population mean, at least when the population is finite. The equation is identical. The only difference is that the symbol for the population mean is μ, the Greek lower-case version of "m." (When the population is infinite, things get much more complicated, because we can't divide by infinity.) In general, population values are symbolized using Greek letters to distinguish them from the sample values.

When to report which type of average

There are dozens of types of averages, of which we have only learned three. This leaves the question of when to use what kind of average in a report. The mean is the best type to use when the distribution is bell-shaped, even if it is not exactly normal. People's heights and weights are good examples, so long as the people are all of the same sex.

The median is a better average when the distribution is skewed, either to the right or to the left. A classic example is income. In the United States, most samples of income are sharply skewed left. If everyone at Microsoft except Bill Gates worked for $10/hour, the mode and median would be $10/ hour, and the mean would be much higher. The mean can be a perfectly dreadful measure of average income.

The mode can be a good measure when the most common values are not near one another in the distribution and the common value is what you want to know. Suppose we are concerned with highway tolls on the Interstate. Tolls are charged per axel. We want to know the average number of axels per vehicle. Out past the city limits, we find lots of cars and big-rig trucks, but very few panel trucks. Most vehicles have two axels or five axels;

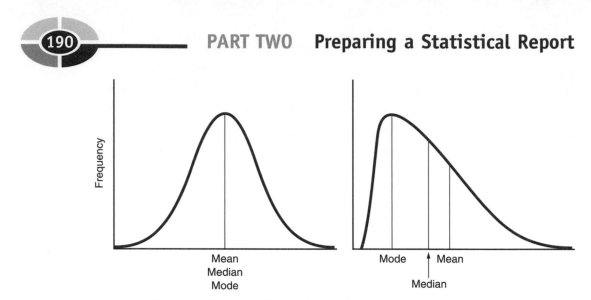

Fig. 8-1. Mean, median, and mode for normal and truncated curves.

a few—panel trucks—have three; none have one. The mean and median might both be close to three, which is deceptive. The mode is two most times of day, and might be five in the middle of the night, when the 18-wheelers are rolling.

The relationship between the mean and the median and the mode for differently shaped distributions is shown in Fig. 8-1. The particular case of the bimodal distribution is shown in Fig. 8-2. As you can see from Figs. 8-1 and 8-2, if we know the mean, the median, and the mode, we know a lot more about our sample than if we just know one type of average. And if we plot the curve for the sample, we know even more.

CRITICAL CAUTION

Bimodal Distribution: Stop the Presses!
If you see a bimodal or multimodal distribution in your data, stop work and take a close look at it. There are very few statistical tests that are meaningful for multimodal distributions. In any event, the mere fact that our data are multimodal almost certainly is a more important and useful fact to know in making our business decisions than we will get from the snazziest statistical test.

If you can examine your data and determine the cause of the bimodal distribution, you may be able to break your data into two workable sets. For example, suppose that your data show a bimodal distribution of the weight of sacks of your product from one plant. You might check with the plant manager and discover that there are two filling machines. If you get the data from each machine, it may be unimodal. Then you can perform statistical tests on the sample data from the output of each machine.

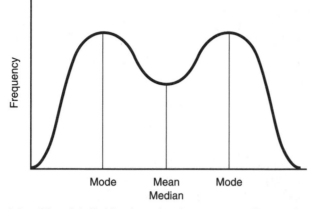

Fig. 8-2.   Bimodal distribution, showing mean, median, and mode.

Averages—especially means and medians, but modes also—have a place in business. A good example would be the manufacture and pricing of various sizes of clothing. Consider men's shirts. If we know the distribution of sizes of men's shirts, we will know how many of each to manufacture. But even if the quantity of sizes purchased falls over a standard distribution curve, where mean, median, and mode are all the same, it would be a mistake to make just one size of shirt! A bigger challenge happens at a small men's store with little storage space. Given the small number of customers, the sample of people who come into the store month-by-month may vary greatly from the population distribution. Is it better to stock a lot of sizes of a few shirt styles and colors, or to have more styles and colors, but not offer as many sizes of some? These problems—which we can describe statistically—are more often solved through trial-and-error and specific business practices than they are through application of statistical analysis. One small shop owner might simply get to know his clients well. Another might handle special orders promptly. Another might open a store especially for tall men. However, at the aggregate level—including manufacturing and purchasing for entire store chains—up-to-date knowledge of the averages and distribution is crucial for business success.

FUN FACTS

One-Size-Fits-All Houses
Huff and Geis (1954) give a wonderful example of the misuse of averages. After World War II, home builders correctly figured out that the average American family was having 2.3 children. They built thousands of housing developments nationwide with just one size of house that was too small for many and too big for many others.

They built for the mean, but their revenue—and their customer's comfort—would have been better served if the variance had been taken into account. A little statistics can be a dangerous thing.

VARIABILITY: THE VARIANCE AND THE STANDARD DEVIATION

The next two statistics, like the range, give us a measure of how spread out the values in the distribution are. In statistics, measures of spread are called measures of *dispersion*. Unlike the range, but like the mean, they are both sufficient statistics, taking in information from all the values in the distribution.

The variance

The variance is a parameter of the normal distribution and also a moment. Technically, it is the second moment about the mean. This means that it is the second moment in the infinite sequence of moments that starts with the mean. This fact tells us a bit about the logic of moments. For its particular definition of typicality, the mean captures all of the information available. Given that we know all we can know about what is a typical value for a distribution, the next logical thing to know would be how the values differ from that typical value. The variance does just that. In a sense, it is a measure of the average distance between each value and the mean. Its formula is:

$$s^2 = \sum (x - m)^2 / (N - 1) \tag{8-5}$$

Equation 8-5 is not as complex as it looks, but it does require a bit of an explanation. The first thing to see is that the symbol for variance is a bit odd in that it is two symbols, an "s" and a "2" that makes the variance look like it should be read as "s-squared," which is a perfectly good thing to call it. While the formula on the right is not the square of anything, we will see in the next section that we define an s—the standard deviation—so that the variance is the square of it.

The next thing to see is that, just like with the mean, we have a sum divided by a count. So, once again, we are working with the "fair share" principle. Inside the sum, we have our new friend, the mean, and it is being subtracted out. This is because the mean is the first moment, and we want the variance, as second moment to include the information *not* included by the mean. The mean is subtracted from each and every value, which leaves us with N numbers that tell us how atypical each original value is.

Now, we come to the little number "2" which, of course, means "squared." The squaring looks complicated, but it is just for convenience. Remember that when we square a negative number, we get a positive number. $(-3) \times (-3) = (-3)^2 = 9$. In the equation for the variance, we square the difference between the value and the mean so that all of the differences are positive. (We could have used some other function, like the absolute value function, to make everything positive, but the absolute value function tends to make pretty curves, like the bell-curve, into very ugly ones, at least from a mathematical point of view.)

And why do we need all the differences to be positive? First, we are going to get a mix of negative and positive numbers. Every value that is less than the mean will be negative when the mean is subtracted from it. All of the values greater than the mean will be positive. If we were just to add all these differences, they would tend to cancel out and give us zero. We would lose the information about how much the data are spread out away from the mean overall.

The sum at the top of our fraction will add up all of the spread of each value of the mean (modified somewhat by the squaring). When we divide by the count, we get a measure of how much each unit varies from the mean, independent of the sample size.

Note that, in this case, we don't divide by the exact sample size, N. Instead, we divide by one less than the sample size. This has to do with the degrees of freedom. In the case of the mean, the total amount of information depends on the sample size, so we divide by the sample size to get one consistent value for the population. In the case of the variance, we have already used up a small amount of information from our sample when we calculated the mean. Because we subtract out the mean from the total at the top of the fraction, the total amount of information is based on one less number, so we divide by $N-1$. This ensures that the mean and the variance are both scaled to the same size, corresponding to one individual unit.

KEY POINT

The variance for the population, when the population is finite, is based on N, not $N-1$.

The standard deviation

The variance has several mathematical advantages. However, because of the squaring, it is no longer scaled to the same size as the original differences.

As we saw in Chapter 3 "What Is Probability?" in our discussion of probability and the normal curve, the differences between numbers on the number line can be extremely valuable in calculating probabilities from distributions. We need a measure of the spread of the distribution that is scaled like a difference. The solution is simple. In Equation 8-6, we just take the square root of the variance and we get exactly the measure we need. It is called the *standard deviation* and it is symbolized with an *s*. Here is how it is defined:

$$s = \sqrt{s^2} \tag{8-6}$$

The standard deviation acts like a unit of measurement, like the inch or the pound. No matter what the scale of the original measurement, so long as the distribution of our measures is bell-shaped, the standard deviation marks out specific areas under the bell curve, starting from the mean at the center of the distribution. And, as we learned in Chapter 3, each area under the bell curve stands for a specific probability that our measurements will have that value. We describe differences along the number line as being "one standard deviation," or "three and a half standard deviations," etc. From the standard normal curve (which has a mean of zero and a standard deviation of one), we know exactly what proportion of the values will fall between any two points measured in standard deviations.

As always, we are really concerned with things in the world, not just in our study, so the standard deviation we are concerned with is the standard deviation of the population, not of the sample. The standard deviation of the population is symbolized with a Greek letter, σ, pronounced "sigma."

HANDY HINTS

The normal curve with all values translated as in Fig. 8-3 so that the mean is zero and the standard deviation is one is called the *standard* normal curve. The values for a standard normal curve are called *z*-scores. Equation 8-7 shows the formula for translating the value of any normally distributed variable into a *z*-score:

$$z = (x - \mu)/\sigma \tag{8-7}$$

where μ is the population mean and σ is the population standard deviation. The probability associated with any *z*-score can be calculated using a computer, or a printed numerical table of the kind often still provided in statistics texts.

In Fig. 8-3, we see a translation from standard deviations into probabilities. Each vertical line is exactly one, two, or three standard

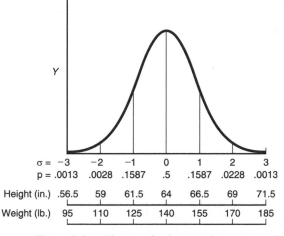

$\sigma =$	−3	−2	−1	0	1	2	3
$p =$	.0013	.0028	.1587	.5	.1587	.0228	.0013
Height (in.)	.56.5	59	61.5	64	66.5	69	71.5
Weight (lb.)	95	110	125	140	155	170	185

Figure 8-3. The standard normal curve.

deviations (called 1-, 2-, or 3-sigma) above or below the mean. The numbers below each line give the precise probability of a value falling *further* from the mean (at 0-sigma) than that line. From these standard points above and below the mean, we can calculate the probability of any value occurring. We have added two additional x-axes below the curve, one for the heights of women in their 20s (like Judy's friends) and another for their weights. These are merely examples to show that no matter what the original measurements, we can use standard deviations to link the values to probabilities.

When we want to talk about values very much larger or very much smaller than the mean, we measure in units of standard deviation from the mean. We speak of "five standard deviations below the mean," or "three standard deviations above the mean." This refers to the probability of being *further* from the mean than that vertical line. Under a normal curve, the probability of being either three standard deviations above or below the mean is just about one in a hundred. That is where the expression, "more than three sigma" comes from. It is just statisticalese for "one in a hundred." In Part Four, Chapter 17 "Quality Management," we will see a discussion of "six sigma," which means that we are looking for items beyond six sigma above, or six sigma below, the mean. Those are very rare events, so six-sigma quality is very high quality.

Measuring Measurement

Early on, we said that statistics not only allows us to measure things while controlling error, but that it also allowed us to measure the error. This involves taking statistics that measure other statistics. In addition, statistics can measure relations between two different variables as well as characterizing the distribution of one variable.

ERRORS AND DEVIATIONS

If we want statistics that measure error, we need to understand about error measurement. And that means we have to understand about measurement error. Measurement error is the errors that occur in measuring things. Error measurement is the measurement of measurement errors. An example (adapted from Huff and Geis, 1954) will make this clear:

In the days before GPS and lasers, people used to measure distances by pacing them off. A person would measure her pace (two normal steps) from heel mark to heel mark, and then walk along a fence or a property line, counting her own paces. Twenty 5-foot paces would mean a distance of 100 feet, sixty would be 100 yards, etc.

Of course, walking over uneven terrain, without military training (where every step is intended to be the same length), pacing is not as accurate as a laser, or even a GPS. Suppose one person paces off 100 yards, based upon their own pace length, and then measures the distance paced more exactly, with a laser land surveying instrument. The exact length of the distance paced would be approximately 100 yards, but almost certainly not exactly 100 yards. Let's say it is 100.837 yards. The difference between the distance we paced off and 100 yards, +0.837 yards, is the measurement error.

We repeat the procedure, and this time we get 99.128 yards. (Our measurement error is −0.872 yards.) We repeat the procedure 10 or 20 times. By this time, we are getting tired, but we also have 10 or 20 numbers representing our attempts to measure off exactly 100 yards using the measurement technique called "pacing." We subtract 100 from each number and we get the measurement error for each attempt. These data represent how far off we are when we try to pace out exactly 100 yards. The limit to our exactness is based on our measurement technique, pacing. If we can take these data and summarize them into a single number, that number will be the error measurement for that particular technique. When we use one measuring method (the laser) to check repeated measurements of another method (pacing), we are *calibrating* the second method against the first. Calibration is the way we determine how accurate our measurement methods are.

Error measures: the probable error and the standard error

The collection of measurement errors is, of course, a distribution, and we can characterize it in any number of ways. For instance, we could use a measure of dispersion, like the variance. The problem with this is that the variance is a measure of dispersion about the mean. Ideally, we would like it if the mean of all of our measurements was the exact true value, but we cannot be assured of that. Recall from our discussion of reliability and validity that, independent of how close we come to the target, there is also the issue of whether or not our attempts center around the target. Do we fall short as often as we go long? If not, then our measurements are biased. We need a measure of dispersion around the target, whether or not that target is at the center of our attempts.

The two most common measures of error are the *probable error* and the *standard error*. The probable error is the distance above or below the true value of what we are trying to measure where half of our measurements will fall. So, if we find that, half of the time, when we try to pace out exactly 100 yards, the distance we actually pace out falls between 98 and 102 yards, then our probable error is $\pm 2\%$. (We use percentage here because everything is measured out in proportions.) If we then walk out and pace a fence line and find that, according to our measurement, it is 150 yards long, we should record that measurement as 150 ± 3 yards (because 2% of 150 is 3). We add the plus-or-minus sign to indicate that we know that our pacing is not a perfectly accurate measurement.

TIPS ON TERMS

Error bounds. The values above and below a measurement showing the uncertainty of the measurement due to measurement error. Usually indicated with the plus-or-minus ($\pm$) symbol.

The plus-or-minus notation assumes that our measurement is unbiased. If we had found out that, when we paced, we tended to overestimate or underestimate the 100-yard standard, then the real plus and minus would not be equal. We might have found that half of our measurements fell between 99 and 103 yards. Biased measurements have skewed distributions, not bell-shaped ones. In order to use error measures like the probable error and the standard error, we need bell-shaped curves when we calibrate. In measurement, we have to accept some error, but we want to remove all bias, if we can. If our measurements are truly unbiased, the mean of our calibration measurements will be 100 yards.

The standard error is based on the assumption that not only is our measurement method unbiased, but that the distribution of errors is truly normal (bell-shaped). We can check this when we calibrate. We measure the same thing over and over again and we check the shape of our distribution to make sure that the mean of our measurements is 100 yards and that the distribution is bell-shaped. As we saw earlier, every set of numbers that is normally distributed can be translated into probabilities using the standard deviation. The standard error of measurement is just the standard deviation applied to the distribution of errors found in calibration, instead of to the distribution of measurements for a sample from a population. Instead of listing the values around the mean where half of our measurements fall, we use the values corresponding to *plus or minus one sigma*. As we can see from Fig. 8-3, .6826, or about two-thirds of our measurements, will fall between these values. (.1587 of our measurements fall below this range and .1587 fall above it. And $1 - (.1587 \times 2) = .6826$.)

CRITICAL CAUTION

It is very important to distinguish between the ordinary standard deviation and the standard error. Take the example of the heights of Judy's friends. Judy's friends are a sample of women in their 20s. When we measure each woman once, the distribution of heights will have a standard deviation of about $2\frac{1}{2}$ inches. We would have measurements for women as short as $5'1''$ and as tall as $5'9''$ in our sample.

On the other hand, if we are calibrating our method for measuring people's heights, we would measure just one woman, let's say Caroline, over and over again. Since Caroline is $5'3''$, almost all of our measurements would be close to 63 inches. In other words, the spread of the distribution of measures of Caroline's height will be much less than the spread of the distribution of measures of the heights of all of Judy's female friends. The standard error of measurement is always much smaller than the standard deviation of the sample or the population.

We can see an example of different ranges of acceptable error in the field of project management. Early in a project, our first estimate of time and cost—before we know many details of what we are doing, or how we are doing it—is called an *order of magnitude estimate*, and should be in the range of -25% to $+75\%$ of the actual project cost. Later, as we refine our project plan, we prepare a *budget estimate*, in the range of -10% to $+25\%$. A *definitive estimate* meets the more demanding standard of -5% to $+10\%$, and is used only when project activities are tightly controlled. The asymmetrical error ranges indicate that a good estimation process, over time, will have actual results fall somewhat below the halfway point of the range of the estimate. If we track estimated and actual project results in detail, we create a historical record of information about our estimated and actual work effort. These data can be used for fine-tuning our estimation procedure, that is, to calibrate it.

Sampling distributions: the standard error of the mean

There is a big problem in using the standard error of measurement to describe our measurement error. In order to determine the standard error of measurement, we need to be able to calibrate our measurement method against a more precise and accurate method (like the laser for distance). This is useful for the measurement of physical quantities, where the most accurate method may be too expensive or cumbersome to use in the field. But, for many measurements, particularly psychological measurements, we may be using the most accurate measurement we have.

For example, how are we to calibrate our measurements of I.Q. for prospective job candidates? We can give the same I.Q. test over and over again to one person, and we will get slightly different results each time, and the distribution will be bell-shaped, but there is no magic ruler we can apply to this person to determine her "true" I.Q. It is as if we had a long fence and we paced it 20 times, but we had no laser or measuring tape to determine the exact length of the fence. We would know how long, in paces, the fence seemed to be each time we measured it, but we could not subtract off the exact length of the fence from those measurements. We have only one measurement method, pacing, and only one scale, paces.

Of course, if we had a short measuring tape, we could translate the length of our pace into feet, even if we could not measure the entire fence. For example, if we knew our pace was exactly five feet long, we could translate each of our measurements of the fence into yards, just as before. But we would still have 20 different measurements of the fence, and no way to know the true length. (There is an old saying that a person with a watch always knows the time, but a person with two watches is never sure.)

In order to solve this problem, statisticians make two critical assumptions about measurement error:

- measurement error is unbiased
- measurement error is normally distributed

Only the second of these assumptions can be tested, and then only to a certain degree. We can measure and re-measure one individual subject. If the distribution of the measurements is not bell-shaped, we know that the distribution is not normal.

With these two assumptions, we can estimate the exact, true value of the variable we are measuring. We measure the one individual over and over again. If the measurements are unbiased, then the true value will fall in the middle of the distribution of measurements. If the distribution is normal, then the best estimate of the central tendency of the distribution is the mean. So our best estimate of the true value of the variable is simply the mean of the measurements taken.

Of course, the "true value" is just an abstraction. Suppose we measure Caroline's height over and over again, and take the mean of all the measurements. Because Caroline is 5′3″, the mean will come out very close to 63 inches. We could take thousands of measurements if we liked, and be able to calculate the mean to many, many decimal places. Suppose we did this, and got a mean of 63.003861. Would this mean that Caroline's height was exactly 63.003861 and not, say, 63.003862? Of course not. During the course of a day, human height varies as much as half an inch. Caroline's height changes as her spine compresses, as she shifts her weight, as she breathes. Caroline has no single height, exact to within a tiny fraction of an inch. There is no one "true" value for the variable. Even a steel rod has no exact length, if we consider the molecular motion of the atoms at either end.

So, how many times should we measure the heights of each of Judy's friends before we can be confident we have enough information to calculate the height accurately enough? As it turns out, we only need to measure each height once and we will still be able to list the error bounds based on the exactness of our measurement method. This is due to a trick called a *sampling distribution*.

A sampling distribution is a distribution of statistics, rather than a distribution of individual data values. For example, we have been thinking of Judy and her female friends as a sample from the population of women in their 20s. Instead, we can think of them as the entire population of Judy and her female friends. Suppose we did not have access to all 14 of these women. We might measure a sample of 4 or 5 of them. Any statistic calculated from this sample could be used to estimate the value of that statistic for the population of all 14 women. Let's use the mean as an example.

Because it is an estimate, the mean of our small sample will not be exactly the mean of the whole population. Because the sample is so small, it will not be very close, on average. But suppose we drew a number of samples and took the mean of each. Then we would have a distribution of means (or any other statistic). This distribution would itself have a mean, a standard deviation, and the like. Essentially, a sampling distribution is a way of calculating statistics for statistics. The standard error of a statistic is just the standard deviation of the sampling distribution of that statistic, where each number in the sampling distribution is a statistic based on a separate sample of size, N.

What does the standard error for a statistic tell us? If we take a sub-sample, such as the heights of four of Judy's female friends, and take the mean, we can feel confident that there is about a 68% chance that the mean of the heights of Judy and all her female friends is within the range of the mean we have, plus or minus the standard error of that mean. The larger our sub-sample, the closer the mean of our sub-sample will be to the mean of the population. So, the standard error will get smaller as N gets larger. As it turns out, there is a simple formula (Equation 8-8) for calculating the standard error of the mean if we already know the standard deviation for the entire population.

$$s_m = s/\sqrt{N} \tag{8-8}$$

The standard deviation of the heights of Judy and all her female friends is 2. If we take any four of these women, for example, Hannah, Liora, Maria, and Ng, the standard error of the mean will be 1 inch (2 divided by the square root of 4). The mean of their heights is 65.5 inches, which is only half an inch from the mean of the entire population, which is 65. Even if we did not know the mean of all of the population, we could state with confidence that there was a two-thirds chance that the mean of the heights of Judy and all her female friends was 65.5 ± 1 inches.

The example above was done with the empirical distribution of a small population, but the formula works just as well for a theoretical distribution of an infinite population. This is key to making use of the standard error of the mean. Judy and her 13 female friends are a sample ($N = 14$) of the population of women in their 20s. From previous studies of women's heights, we know that the standard deviation of young women's heights is $2\frac{1}{2}$ inches. If Judy does not pick her friends on the basis of their height, we can say with confidence that the mean height of the women in the entire population is 65 (the mean of the sample) plus-or-minus .67 (2.5 divided by the square root of 14).

EXERCISE

A brief exercise will illustrate how the standard deviation, the standard error of measurement, and the standard error of the mean, differ from one another and are all useful in different ways.

Suppose a particular department is having trouble getting effective performance out of their employees. The supervisors blame the training program for not giving effective training. The trainers blame the department for not hiring people smart enough to learn the job. We can't afford to screen all job applicants with an I.Q. test, but we can afford to either (a) hire a training consultant to revise the training program or (b) initiate a recruiting campaign to attract smarter job candidates, but not both. If we can find out the average I.Q. of the job applicant pool, we can decide how to spend our money.

We can't calculate the average I.Q. for the entire job applicant pool. We don't even know who they are. The job applicant pool is all the folks who will eventually, some day, apply for this sort of job with us. However, we do know that the standard deviation of I.Q. scores across many different populations is 15 points. We can pay a small number of our current job applicants (say, 16) to take an I.Q. test while they are waiting for their interviews. Whatever the mean I.Q. we get for these 16 applicants, we can state with confidence that the mean for the entire job applicant pool will be that number plus or minus 3.75 with a 68% probability. As an exercise, explain why this is so.

Another way to assess the effectiveness of the training program is to see if the problem employees in fact have low I.Q.s. Suppose we are informed that folks with an I.Q. below 110 cannot be expected to perform this job effectively, no matter what the training. We decide to re-assign any problem employees with I.Q.s below 110, which, unfortunately, means a reduction in pay. George works at this job and has had problems. However, he is highly motivated and is willing to take additional training. His I.Q. is 108. The standard error of measurement for this I.Q. test is ±3 points. From a statistical point of view, we should retrain George, not reassign him. For your exercise, explain this. (Note that we do *not* need to test George over and over again. The standard error of measurement for the *test* tells us how far off any score may be for any person tested.)

HANDY HINTS

There is another way of thinking about these sorts of questions that does not involve error bounds directly. Suppose we want to know if our applicant pool for the jobs in the exercise above have the same I.Q. as the general population. We know that the I.Q. of the general population has a mean of 100 with a standard deviation of 15. We measure the I.Q.s of our sample of 16 job applicants and find that they have a mean I.Q. of 105. We can create a standardized z-score from the sample mean, based on the population mean and the standard error of the mean. This z-value can be translated into a probability that a sample mean could be that far from the population mean solely by random chance.

The formula is in Equation 8-9:

$$z = \frac{(m - \mu)}{\sigma}$$

(8-9)

where μ is the population mean and σ is the standard error of the mean calculated based on the population standard deviation.

For our example, this means the z-score is $(105-100)/3.75 = 1.33$. If the mean I.Q. of applicant pool is really 100, the likelihood of getting a sample mean five points or more away from 100 is .1836 or less than one in five. There is a better than 80% chance that our job applicant pool has a higher than average I.Q.

We will learn more about this way of thinking about sample means when we discuss *significance testing*.

RELATEDNESS: THE CORRELATION COEFFICIENT

As we saw in Chapter 7 "Graphs and Charts," we can examine the relationship between the distribution of two different variables for the same sample with a scatter plot. The correlation coefficient is a statistic that summarizes the relationship we can see in a scatter plot in a single number. The equation for the correlation coefficient (called the Pearson product-moment correlation to distinguish it from other measures of correlation) is:

$$r = \frac{N(\sum xy) - (\sum x)(\sum y)}{\left(\sqrt{N\sum x^2 - (\sum x)^2}\right)\left(\sqrt{N\sum y^2 - (\sum y)^2}\right)} \tag{8-10}$$

There are a number of different forms of Equation 8-10, but this one can be explained easily. The two elements under the square root signs in the denominator are measures of the variability of each of the two variables. Note the similarity of the numerator to the two parts of the denominator. The numerator is a measure of how much the two variables vary together, or *covary*. The covariance is a measure of how much one variable goes up above the mean (or down below the mean) when the other variable goes above or below the mean.

The Pearson correlation coefficient is a ratio of a measure of the covariance to the total variability of both variables. It ranges from -1.0 to $+1.0$. A correlation of $+1.0$ means that however much the value of x differs from the mean, the value of y differs exactly proportionately. The covariance is exactly the same as total variance of both variables. In a scatter plot, all of the points would lie on a straight line going from the lower left to the upper right. (This is sometimes called a *perfect correlation*.) A correlation of -1.0 means that however much the value of x differs from the mean, the value of y differs exactly proportionately, but in the opposite direction. When *x* is above the mean, *y* is below it, and vice versa. In a scatter plot, all of the points would lie on a straight line going from the upper left to the lower right. A correlation of 0 means that the two variables are completely unrelated. There is no pattern relating the variability of one variable and the other. In a scatter plot, all of the points would be scattered uniformly in a filled circle.

There are many uses of correlation in business. We can search for correlations in the market—do customers who purchase one product tend to like certain other products? This sort of information can guide our marketing efforts with current customers. We will hear more about correlation when we learn about regression, which is a statistical way of looking at cause and effect using the values of individual subject units.

VARIABILITY OF A DIFFERENCE: THE *t* STATISTIC

The first key to statistical inference is the ability to translate numerical values into the probability that those values occur. This is accomplished by translating the values into *z*-scores. However, there is a problem with this method. We can only calculate a *z*-score if the standard deviation of our sample is the same as the standard deviation of the population. If the two standard deviations are different, we do not know which one to use to calculate the *z*-scores.

Suppose that, in the earlier example, we are not only unsure that our job applicant pool have an average I.Q. of 100, but we are also unsure that the standard deviation of their I.Q.s is 15 (as it is for the general population). We could, of course, calculate the sample standard deviation from our sample of 16 applicants, but this uses up some of the information from our sample. (Remember our discussion of degrees of freedom!) The larger our sample, of course, the better our estimate of the mean will be and the lower the proportion of information spent on calculating the sample standard deviation.

The normal curve is based on an infinite population. There is no way to include the size of the population (or the sample) in calculating probabilities from *z*-scores. However, there is another bell-shaped curve, called Student's *t* distribution, that does change shape depending on *N*. When *N* gets very large, Student's *t* distribution approximates the normal distribution. With lower values of *N*, it is shorter and stubbier.

We can calculate probability values from a *t* distribution just as we can from a normal distribution (either using a computer or a printed numerical table), by using Equation 8-11. The standard score translated from the

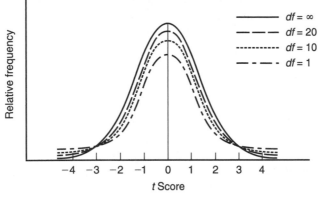

Fig. 8-4. *t*-Score distributions.

sample mean of a sample of size *n* with sample standard deviation, *s* (not σ) is:

$$t = \frac{\bar{X} - \mu}{s_x/\sqrt{n}} \tag{8-11}$$

where *n* is the population mean we are comparing the sample mean to.

Unlike *z*-scores, the calculation above only works for means and not individual scores, because there is no way to calculate the standard deviation for a sample of size $n = 1$. This limitation is offset by the fact that, unlike the normal curve, the *t* distribution can be used to compare two sample means when the population mean is unknown. We will hear more about this use of the *t* statistic when we learn about *hypothesis testing*, which is a statistical way of looking at cause and effect using summary values from groups of subject units.

There are many other statistics used for many other purposes. We will see a small sample of these in the context of different statistical techniques we will learn about in Part Three.

Quiz

1. The _____ of all the numbers in a group of numbers is the basis for almost every other statistic.
    (a) Ratio
    (b) Count
    (c) Frequency
    (d) Proportion

2. _____ is taking a statistical measure of a sample and inferring the value of that same measure for the entire population.
   (a)   Ratio
   (b)   Count
   (c)   Estimation
   (d)   Proportion

3. The minimum and maximum values in the sample distribution tell us the _____.
   (a)   Range
   (b)   Estimation
   (c)   Count
   (d)   Ratio

4. The _____ is a type of average.
   (a)   Mean
   (b)   Median
   (c)   Mode
   (d)   All of the above

5. The _____ is the square root of the variance.
   (a)   Mean
   (b)   Normal curve
   (c)   Standard deviation
   (d)   Error

6. The _____ is the distance above or below the true value of the measure where half of our measurements will fall.
   (a)   Probable error
   (b)   Standard error
   (c)   Standard deviation
   (d)   Variance

7. The _____ is the standard deviation applied to the distribution of errors found in the calibration.
   (a)   Probable error
   (b)   Standard error or measurement
   (c)   Variance
   (d)   Distribution

8. If we can assume that measurement error is _____ , we can estimate the exact true value of the variable we are measuring.
   (a) Unbiased
   (b) Normally distributed
   (c) Both (a) and (b)
   (d) Neither (a) nor (b)

9. The range of a correlation coefficient is . . .
   (a) 0 to +1
   (b) −1 to 0
   (c) 1 to 2
   (d) −1 to +1

10. The _____ can be used to compare two sample means when the population mean is unknown.
   (a) *t*-distribution
   (b) Correlation coefficient
   (c) Standard deviation
   (d) Standard error

A Difference That Makes a Difference. When Do Statistics Mean Something?

In Chapter 1 "Statistics for Business," we defined inferential statistics as statistics that allowed us to draw conclusions (find answers to specific questions) from data. The hallmark of inferential statistics is the notion of *statistical significance*. To say that something is statistically significant is really to say two things: First, enough evidence has been put together to state a conclusion with confidence. Second, the evidence taken from the data has leaned far enough one way or the other to give a decisive answer to our question. We will examine both parts of statistical significance later on in this chapter.

In a business context, the most important thing to know about statistical significance is that the rules for saying how much evidence is enough, either to be confident or decisive, were developed for the social sciences, not for business. The kinds of questions asked in science can be very different from those asked in business. Scientific questions come from the overall goals of the scientific enterprise, which are very different from the goals of business. As we will see, the specific statistical measures that tell us when we can be confident and decisive do not always measure up to the requirements of business decisions. We need to be very careful in adapting the scientific notion of statistical significance to answering questions that arise in the context of making business decisions.

The Scientific Approach

The goal of science is to find out about how the world works, mostly by asking questions. The goals of business are to make a sustainable profit, to grow the business, to expand market share, and so forth. In business, we can often make use of the answers to questions about the world in making decisions that improve our chances of achieving business goals. However, we need to understand exactly how what makes for a good scientific answer relates to what makes for a good answer to a business question. For that, we need to understand how the notion of statistical significance relates to what makes for a good scientific answer.

A LITTLE HISTORY OF SCIENCE

Round about 1926, the famous scientist and mathematician, Sir Ronald A. Fisher, laid out the rules for statistical significance. Fisher believed that science was about discovering causes. He developed several principles of statistical inference that led to the definition of statistical significance as the criterion for accepting a statistical conclusion in science. In his view, scientific progress was all about abandoning old theories for new, better theories. (This was a popular view back then.) The question for science was how much evidence for the new theory is enough to abandon the old theory.

The big problem with statistical inference is that even if we have all our facts right, there is a possibility that our conclusion will be wrong. In the context of science, we might abandon a good theory for a bad one. According to Fisher's view of scientific progress, this was the worst possible outcome. So Fisher invented the principle of *conservatism*, which can be stated as: *Place an upper limit on the probability of wrongly deciding in favor of the novel theory*. The scientific community would decide on a standard and would agree always to collect enough evidence in favor of any new theory to a point where the likelihood that the new theory was wrong dropped below that standard point. Only then would the new theory be accepted. Fisher called the standard the α-*level* (pronounced "alpha-level") and suggested a ratio of 1/20, or $\alpha = .05$.

In the social sciences, this inferential strategy has proved remarkably durable. Despite many problems over the decades, it remains the dominant method for using inferential statistics across the social sciences, both basic and applied.

CAPITALIZING ON CHANCE

The key to using an α-level is understanding another of Fisher's concepts, *capitalizing on chance*. Suppose the current theory says that a particular variable will have a particular value and the new theory says it will have a different value. We collect data and estimate the value of this variable using our statistical techniques. As we learned in Chapter 8 "Common Statistical Measures," the further the mean of our data from the value predicted by the current theory, the more likely it is that the current theory is wrong.

Now suppose that the current theory is correct. That means that the true value of the variable is exactly as predicted. The difference between the mean of our data and that value is due to some combination of sampling error and measurement error. Even if we assume that our measurements are perfect (which they never are), there is some error due to the fact that we can only afford to measure a sample of our population. The reason that the mean of our data is not exactly what the current theory predicts is because we just happened to select subject units with slightly higher or slightly lower values on the variable in question. For example, we might have selected 20 women to find out their height. The likelihood that the mean height for this group is *exactly* the mean height of all American women in that age group is vanishingly small. All statistics assures us is that the mean will be close to the true value so long as our sample is big enough.

Fisher wants to use this statistical assurance as the basis for statistical inference. His idea is to disregard any small difference between the predicted value and the mean of the data, because this is most likely due to either measurement error or sampling error. Were we to take this small difference as evidence that the current theory is wrong and the new theory is right, we would be capitalizing on chance. That is, we would be relying on the luck of having drawn a sample that favored the new theory to wrongly choose the new theory over the old one. Fisher's rule says to move to a new theory only when the odds are at least 19:1 that the evidence for the new theory is due to actual effects and not luck.

CONFIDENCE AND DECISIVENESS

There are a number of problems with Fisher's approach. The most important one for our purposes is that Fisher's criterion is based on avoiding coming to a false conclusion when the evidence is weak, rather than coming to a true conclusion when the evidence is strong. In order to understand this problem, we need to understand the difference between a conclusion we can state with confidence and a conclusion that is decisive.

The larger our sample, the smaller the error of the mean will be. Whatever result we get, we can be confident that that value is close to the true value. Confidence comes with large samples. Fisher's strategy addresses this problem by requiring an α-level. We must take a large enough sample to lower the probability of capitalizing on chance.

If we have two theories that predict different values for a variable, even if we take a huge sample and get a very good estimate of the value, it still might fall halfway between the two values. We would be able to say with confidence that our data did not decide between the two theories. This might be satisfying in an abstract sense, but it would not be very practical for decision-making. Fisher's strategy addresses this problem by biasing the entire inferential strategy against the new theory. If our data aren't strong enough on which to base a decision, we must decide in favor of the status quo. As we will see further on, the entire mathematical structure of Fisherian statistical inference is asymmetrical in order to create this bias.

Now, suppose that the two values predicted by the two theories are very close. We collect enough data to be confident that, if the new theory is wrong, the odds of abandoning the old theory are below the α-level. But this does not guarantee that if the new theory is correct, we will have a very good chance of deciding in its favor. The Fisherian strategy only protects us from falsely abandoning the current theory, not from falsely rejecting the new theory.

In principle, we can take an even bigger sample. A big enough sample will assure us that, if the new theory is correct, we will choose it over the old one, despite the bias in the system. In practice, bigger samples cost more money and resources. The bias in the system makes changing theories expensive. In science, cost is not supposed to be an issue. In business, cost is a very big issue. The goals of business are always affected by cost.

Hypothesis Testing

In the preceding section, we mentioned that Fisher believed that science was about discovering causes. It should go without saying that the information needed to make a business decision may very often not be information about a cause–effect relation. After all, it is a lot more important to know that 90% of women between age 19 and 34 want to buy your new product than it is to know precisely what caused that fact. It should go without saying, but, unfortunately, it does not. Much of statistics comes from work in the sciences, and, in particular, the social sciences, where understanding cause–effect relations is taken to be of utmost importance. Because of this, statistics texts often spend a great deal of time focused on techniques for establishing cause–effect relations without even explaining why cause–effect relations are important, much less taking the time to consider when, in business, other sorts of statistics providing other sorts of information, may be more important.

One strategy in business statistics is to re-cast our decisions so that the information we need is expressed in terms of questions about causes and effects. This strategy may be awkward, but it also ignores the bias in the system. Fisher's view of science involved a view of progress. In Fisher's view, science has to move forward from one theory to a better theory. Business is not about progress; it is about profit and loss. We need to bias our inferential system toward profit and away from loss, not toward the status quo and away from new ideas. The remainder of this chapter will show how the Fisherian strategy works so that we can apply it usefully in business.

THE ROLE OF EXPERIMENTS

In order to connect his inferential strategy to the detection of causes, Fisher had to formalize a particular sort of experiment that would generate the right sort of data; data that could be analyzed according to his rules for good statistical inference. Fisher established a basic design for experiments that could be analyzed in terms of statistical significance. This design was based on Gossett's experiment that illustrated the first use of the *t* test. Because it involved comparisons of groups of subjects, the general term for this new type of statistical analysis is *group tests*.

Fisher proposed a great variety of group test designs, and, in the decades since, applied statisticians working in various fields have greatly expanded the variety of experimental designs that will allow statistical significance to be determined. (It should be noted that there are many types of experiments, including experiments that have won Nobel prizes, that do not fit in with

Fisher's design, and where ascertaining statistical significance is impossible.) We will learn about a number of the types of experiments that permit statistical significance to be calculated in Part Three. In the remainder of this section, we will examine one experimental design, the two-group test, to see how the notion of causes and effects fits in with statistical significance.

The need for experimental control

The first question we need to answer is why use experiments at all? There are lots of ways to collect data, and, mathematically, we can do the calculations for statistical significance no matter where the data come from. But it is only in the case of experiments (and quasi-experiments) that we can be assured that our calculations for statistical significance will provide Fisher's guarantee of conservatism.

The answer has to do with random sampling, which we discussed earlier in Chapter 2 "What Is Statistics?." Fisher's idea for detecting causes was based on *intervention*. If we suspect that event A is the cause of event B, we deliberately intervene and do A to a subject and see if B happens. The question then is whether or not B would have occurred even if we had not done A. (This is the same question that perplexed John Stuart Mill, and Fisher's answer is based on Mill's.)

For example, if we want to improve sales, we can put our salespeople through training. (We will develop this example throughout this chapter.) How would we design the training as a properly controlled experiment? How would we design it for the best business results? If the best experiment does not lead to the highest, or most immediate return on investment, is the value of the knowledge we gain through the experiment worth the cost of choosing the controlled experiment over the optimal immediate business solution?

TIPS ON TERMS

Recall that events are designated with upper-case Roman letters (A, B, C, etc.) and that a corresponding variable is designated with an upper-case Italic letter (A, B, C, etc.).

The first addition Fisher needed to make to the basic notion of an intervention work statistically, was to introduce the idea of a *group*. If we do A to just one subject, we will not have a large enough sample to calculate a reliable mean value. We need to select N subjects and do A to all of them and then measure the value of the variable, B for each. This will give us useful measures of error and ensure that our sample mean is a good estimate of the true mean. So, for our Fisherian experiment to see if training salespeople increases sales, we will need to train a group of N salespeople.

The next thing Fisher needed to add was the idea (also originally due to Mill) of a *control*. We need another group of subjects. These subjects will *not* receive the intervention. That is, we will create a control group of salespeople who will not receive training. In all other respects, they will be treated exactly the same as our experimental subjects. At the end of the experiment, we will measure the value of B for all subjects in both groups. This will give us a mean value for B where A happened and another value where A did not happen. The difference between these two mean values of B will be our experimental measure of the causal relation for the effect of A on B. For the minimum group test, we need two groups, the *experimental group* and the *control group*.

TIPS ON TERMS

Dependent variable. The variable whose values are used in the calculation of the results of the experiment. The value of this variable is intended to measure the effect of the experimental intervention. Also called the **outcome measure**.

Independent variable. The variable whose values are used to define the different groups in a group test experiment. The value of this variable is intended to measure whether or not, or to what degree, the experimental intervention is applied to the subjects. In more complex experiments, there may be multiple independent variables.

For the basic two-group experiment, the measure of the effect of our intervention will be the difference between the two group means. As we mentioned in Chapter 8, a difference between means can be converted into a probability value using the t distribution. As we will see in Chapter 13, where we discuss the statistical tests for group designs, this is the basis of the t test, the first method of statistical inference, which was developed by Gossett (writing under the name of Student).

KEY POINT

The key to understanding the need for experimental controls is found in the notion of a random sample. Suppose we were to perform a two-group experiment and the subjects in our experimental group had higher B values to begin with. At the end of our experiment, we might find a higher mean value for B in the experimental group than in the control group, not due to our intervention, but due to the random chance that our sample was biased. This would violate the principle of conservatism, because we would claim that a causal relationship existed solely due to chance affecting the results of our experiment.

Fisher's solution is the *random assignment to groups*. First, we sample from our population, and then, for each subject, we flip a coin (or the equivalent) to decide whether that subject gets assigned to the control group or to the experimental group. Random chance will ensure that, in the long run, there will be no systematic differences between the groups on any variable, except for those caused by our intervention. Of course, for any particular sample, there is a chance that our random assignment will produce a spurious difference in B values between the groups, but Fisher's design ensures that that chance will not exceed the α-level.

The *P*-value for a hypothesis test

In order to make the results of our experiment fit in with Fisher's principles, the results need to be translated into a probability that expresses the likelihood that the results are due to chance alone. This is the final step in Fisher's scheme. By using our trick for converting a statistical value into a probability, we translate the numerical value of our experimental results into a probability value expressing the likelihood that those results are due to chance. Using a Fisherian design, the stronger our results (relative to the error variance, of course), the *less* likely those results are due to chance. This probability value is called the *P-value*. If the *P*-value is lower than the α-level, then the results are strong enough to reject the old theory in favor of the new. (The *P*-value goes down as the outcome measure goes up.)

In the case of the two-group test, the results of our experiment must be measured as the difference in means between the two groups. This difference will increase as the size of the effect of our intervention increases. This means that the role of the "new" theory is filled by the claim that the causal relationship does exist, that A does cause B. Other experimental designs use different outcome measures, often very complex ones, but the same rules apply. A larger value for the outcome measure always makes for a statistically significant result. And the outcome measure increases as the effect of our intervention increases. Significance means that A did cause B.

FUN FACTS

You may have noticed an odd fact about Fisher's approach. Significance means that the causal relationship exists. Significance also means that the old theory can be rejected for the new. That means that the new theory must predict at least one causal relationship not predicted by the old theory. Such a situation is not uncommon in science, but there are certainly cases, including famous Nobel prize-winning cases, where the discovery was that A did not cause B.

In the social sciences—particularly in psychology, where hypothesis testing dominates experimental work—the various limitations the types of theories that can be tested have had interesting effects. In some areas, theories have been replaced by hypotheses. Our understanding in these fields amounts to a growing list of (somewhat disconnected) statements of causal relations. In other areas, elaborate theories are constructed and compared with regard to a small number of claims where a new theory predicts a causal relation that the current theory does not. This tends to lead to more and more complex theories that only explain a little more data.

The most profound problem created by this strict adherence to Fisher's strategy has been that, if, due to that 1/20 chance, a causal relation that does not really exist comes to be accepted by science, it is extremely difficult to prove that it does not. It is always difficult to prove a negative, but Fisher's strategy makes it much harder in the social sciences. Recently, in psychology, there have been some efforts to eliminate some false theories that have been around for decades. It has been an uphill battle in all cases.

Fisher's model has some limitations in business situations. We will look at these in more detail, returning to our case study of sales training throughout this chapter. But one limitation is important to mention now: In business, there may be no current theory. There is no prevailing theory in business as to whether sales training will increase sales. Some folks—biased folks, like the ones who are trying to sell us the training, and unbiased folks, like researchers in our own company—may hold that they believe it will help. Others—both biased and unbiased—may believe that sales training does not help. But neither theory is the current or prevailing theory. They are just two theories—or perhaps more accurately, hypotheses—to be tested in the experiment. As a result, Fisher's model, which includes a built-in bias toward one theory over the other, does not apply. It would be a mistake to assign either theory to the role of the current theory.

In the simplest case, we can consider our current business practices to be the accepted view, because almost any change will involve additional expenditures (such as the costs of implementing a new training program). This works out well if our business question is whether a change in our current practices will cause an improvement in return on investment (ROI) or some other positive business result. Here, the analogy to the scientific context is sound. We don't want to change our current business practices unless we can be confident that the intervention will make a worthwhile difference.

Businesses are conservative for different reasons than science, but Fisher's logic can still apply. In this context, the only real limitation to this approach is that there is no reasonable way to set the α-level. Later on in this chapter, we will see a better approach that allows us to decide just how conservative to be in making a specific business decision.

The null hypothesis

While Fisher's general inferential strategy was modeled on the t test, it can be expressed much more generally in terms of what Fisher called "hypotheses." Call the outcome variable Y. (In the case of the two-group experiment, the outcome is the mean difference between the two groups. For other types of experimental or quasi-experimental designs, the outcome variable may be something else entirely, so long as it increases with the increasing effect of our experimental intervention.) We assume that the current theory predicts a specific value for the mean of the outcome variable. Call that predicted value, μ_0. Fisher called a prediction of the value of the outcome variable a *hypothesis* and called the prediction made by the current theory, the *null hypothesis*, H_0. Symbolically,

$$H_0 : \mu_Y = \mu_0$$

where μ_Y is the population mean for the outcome variable.

FUN FACTS

Fisher's use of the term "hypothesis" is rather unusual in the philosophy of science. Ordinarily, "hypothesis" means a statement that connects up different facts by explaining some facts in terms of others. Fisher used "hypothesis" to mean a prediction of the outcome of an experiment based on a pre-existing theory.

According to the principle of conservatism, we need to concern ourselves with the case where the current theory is true and the new theory is false. In that case, we can assume that the true value of the mean for the outcome variable is μ_0. Any experiment we perform will result in a measurement value of the outcome variable that differs from μ_0 only due to error. On the other hand, if the new theory is true, then the mean of the outcome variable will have some other value, $\mu_1 \neq \mu_0$. Fisher called this the alternative hypothesis, H_1. So long as we know the distribution of the outcome variable where the null hypothesis is true, we can ensure the principle of conservatism by insisting that the measured value of the outcome variable be far

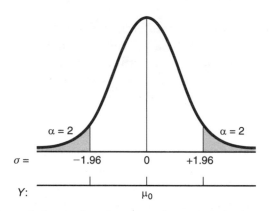

Fig. 9-1. A mean beyond the α-values (gray areas) permits rejection of the null hypothesis where the null hypothesis predicts the mean, μ_0.

enough from μ_0 that the probability of a result that far or further be less than α. Figure 9-1 illustrates this relation.

In Fig. 9-1, the shaded regions are called the "region of rejection," because, if the mean falls in one of those regions, we are justified in rejecting the null hypothesis. Note that we always talk about acceptance or rejection of the null hypothesis and not the alternative hypothesis. Because our criterion is based on the principle of conservatism, we are not truly accepting the alternative. We are merely rejecting the null hypothesis and taking up the alternative tentatively. Every new theory is subject to further testing. This one experiment gives us reason to prefer the alternative hypothesis to the null hypothesis for this one outcome variable. Other experiments will have to be done to solidify this new theory.

HANDY HINTS

Because the α-level is specified as a probability, under a normal curve it always corresponds to one specific value of σ. In Fig. 9-1, we show the σ values for an α-level of .05 for the case where the alternative hypothesis is that the mean is either higher or lower than μ_0. This case is called the two-tailed test, because the rejection region is split between the two tails of the bell curve.

ERROR IN HYPOTHESIS TESTING

We have seen how Fisher's principle of conservatism ties the rules for drawing statistical conclusions to the theory of error. It is only when the

Table 9-1   The four possible results of a statistical inference.

Conclusion	Actual Situation	
	H_0 is True	H_0 is False
Reject H_0	Type I error $p = \alpha$	Correct rejection $p = 1 - \beta$ (power)
Do not reject H_0	Correct acceptance $p = 1 - \alpha$	Type II error $p = \beta$

mean of the outcome variable differs from the prediction of the null hypothesis by more than what error theory would predict that we reject the possibility that no causal relation exists. We have also discussed the fact that rejecting the null hypothesis does not mean that the data show that the alternative hypothesis is more likely than the null, but only that the null hypothesis was less likely than our community standard for acceptance, α. This conundrum can be illustrated by explaining the two types of error possible in statistical inference.

In Table 9-1, we see the four possible situations we can find ourselves in when we make a statistical inference. Because we have defined the alternative hypothesis as any possibility other than the null hypothesis, the only two possible ways the world can be are for either the null or the alternative hypothesis to be true. We can either retain the current theory (called accepting the null) or reject it (called rejecting the null). Accepting the null when it is true, or rejecting it when it is false are the two ways of being correct. Rejecting the null when it is true is called Type I error. Failing to reject the null when it is false is called Type II error.

CRITICAL CAUTION

Note that α is not a name for Type I error. It is the conditional probability of Type I error given that the null hypothesis is true. Likewise β is not a name for Type II error. It is the conditional probability of Type II error given that the null hypothesis is false.

In Fig. 9-2, we see the relationship between the two types of error and the probability distribution of the possible experimental results. We will only

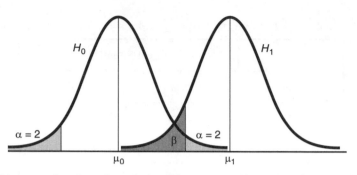

Fig. 9-2. Diagram showing the relationships among Type I and Type II error, and power $(1-\beta)$.

reject the null if the value of the outcome variable is far enough from μ_0 (under the light gray area). Note that β (the dark gray area), the probability of a Type II error, is only partially determined by α, the probability of a Type I error. We base our decision solely upon the light shading of the left-hand curve. If the alternative hypothesis is true, then a Type II error happens when the value of the outcome variable is under the dark area of the right-hand curve. This is when the true value of the outcome variable was far from μ_0, but error variance caused the experimental value to be low enough that we could not reject the null.

Type I error: saying it is, when it ain't

Type I error is when the true value of the outcome variable is μ_0, but the error variance made the experimental results so far from μ_0 that we reject the null. The principle of conservatism directs us to limit the probability of this type of error to a fixed and agreed upon value, called the significance level. This means that, over many experiments, we will only commit Type I error α percent of the time. In the social sciences, this is almost always 5%.

In the scientific formulation, avoiding Type I error is considered paramount, far more important than avoiding Type II error. The reason for this is that the current theory is presumably in place because of previous research. Science can only make progress if it demands a preponderance of the evidence over time before changing any of its views. If science is subject to every popular fad that comes along, it will vacillate so wildly from point to point that any forward progress that would be driven by the data would be overwhelmed. (At least, that was the 20th century view shared by Fisher.) The principle of conservatism is based on the goal of science to build a body of knowledge based on evidence. This long-term goal is rarely of value in making business decisions unless we can tie conservatism to lower cost.

And, of course, distinguishing Type I error from Type II error requires defining one hypothesis as the null hypothesis, which may not make sense in a business context.

Type II error, beta probability, and power of test

Type II error is when the true value of the outcome variable is μ_1, but the error variance made the experimental results so close to μ_0 that we retain the null. When a Type II error occurs, we lose out on an opportunity to improve our view of the world by adopting a better theory. We stick to our old, less correct, theory. In science, this is a low-cost error, because, eventually, as science moves forward and other experiments are tried, the truth will be out. Science is an enterprise without a horizon. The value of keeping on course is far more important than actually reaching our destination at any particular time. The same cannot be said of business, where results must be delivered on schedule.

KEY POINT

Statistical power is a measure of the ability of the statistical test to detect the effectiveness of an intervention. In business, this is often what we are paying for when we request inferential statistics.

CRITICAL CAUTION

The biggest problem with Type II error is that, because the entire process of statistical inference is based around limiting Type I error, the actual probability of a Type II error is hard to calculate. It depends not only upon N, but also upon the difference between μ_0 and μ_1. The closer the two predicted means are, the larger N must be in order to distinguish between the two hypotheses. Figure 9-3 shows a case where the two means are much closer together. The rejection region (light gray) does not change. There is now a larger region (in dark gray) where the experimental results would be closer to μ_1 than μ_0, but we would still reject the alternative hypothesis and retain the null. In fact, the value of the outcome variable could equal the value predicted by the alternative hypothesis exactly and we would still reject the alternative! (We can see this in Fig. 9-3, because μ_1 is underneath the dark gray shaded region.)

All too often, the value of μ_1 is unknown and can only be guessed at. A set of complex calculations, called *power calculations,* which differ depending on the design

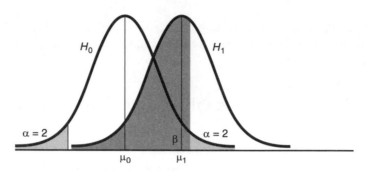
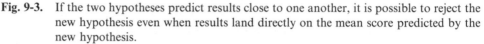

Fig. 9-3. If the two hypotheses predict results close to one another, it is possible to reject the new hypothesis even when results land directly on the mean score predicted by the new hypothesis.

of the experiment, can be used to estimate how large N must be in order to detect a difference between the two predictions as small as some particular value. (Recall from Fig. 8-4 that the larger our N, the narrower our bell curve, and the closer the rejection region will be to μ_0.) Many experiments are performed that fail to reject the currently accepted theory solely because enough data were collected to guarantee conservatism, but not enough data were collected to guarantee power. Increased statistical power means having the rejection region close enough to μ_0 to be able to distinguish between theories that predict very similar means.

STURDY STATISTICS: NONPARAMETRIC TESTS

Up until this point, we have been talking about hypothesis testing in terms of means and errors. However, so long as we can translate our results into a probability that the intervention has had no effect, we can assert that the results are statistically significant whenever that probability is below the significance level. Statisticians have developed dozens of techniques for translating different types of effects on different types of data into measures of statistical significance. These inferential techniques are divided into two broad classes: *parametric* and *nonparametric* tests.

Parametric tests involve sampling distributions of a statistic, usually the mean. The test measures the effect of the intervention in terms of some specified change in that statistic. The translation of that change into a probability value is based on assumptions about the error that produces the sampling distribution for the statistic. The theoretical distributions of these sampling distributions are defined in terms of a few numbers, usually the mean, the variance, and possibly the degrees of freedom. These defining numbers are called parameters, hence the name, parametric tests.

Nonparametric tests, also known as distribution-free tests, is a catch-all term for any other kind of inferential statistical method. The most common thing they share is that they require that the data satisfy fewer assumptions (such as interval level scale, normal distribution, etc.) in order to work. As such, Mosteller and Rourke, in *Sturdy statistics: Nonparametric and order statistics* (1973) have called them "sturdy statistics."

STUDY REVIEW

A Sturdy Statistic is Like an SUV
What makes a four-wheel drive SUV (Sport Utility Vehicle) sturdier than an ordinary small car? It is able to handle rougher roads. In the same way, sturdy statistics are statistical tools that can handle rougher data. If our data do not follow the normal curve, or if it is multi-modal, we are not going to be able to use parametric statistics, because our data do not meet the parameters. If we think of the normal curve as the highway of data, then sturdy, nonparametric statistics let us go off the road—work with data that aren't characterized by simple parameters—and still reach our destination; informative statistical results.

And, like a four-wheel drive vehicle, there is usually a price to pay for being able to traverse such rugged territory. Just as off-road vehicles tend to have lower gas mileage than economy cars and lower top speeds than sports cars, nonparametric statistics tend to have lower statistical power than parametric statistics.

As we will see in Part Three, different parametric tests are designed to handle different types of experiments. So are the different nonparametric tests. In fact, for the most part, there is usually a nonparametric test available for almost any situation in which there is a parametric test available.

When to choose a sturdy statistic

There are a number of differences between parametric and nonparametric tests. The decision as to whether to use a parametric or a nonparametric test depends on understanding the advantages of each.

The traditional trade-off for choosing parametric over nonparametric tests is the need to satisfy statistical assumptions versus the need for statistical power. As we will see in Part Three, parametric tests only work if certain pre-existing conditions are true. The most common of these involve the level of measurement and the distribution of the data. The mathematical proofs that demonstrate that a particular test will generate an accurate *P*-value almost always involve assumptions. Traditional parametric tests are proven to work when the data are measured at, at least an interval level and when the population distribution of the variable is normal. There are usually a few other assumptions. The bottom line is that, unless these assumptions are met, we cannot be assured that the *P*-value will be accurate enough to ensure conservatism. In science, the absence of such a guarantee means that the inference is invalid.

Many parametric tests are *robust* under violations of some (but not all) of these assumptions. This means that the particular assumption has to be dramatically and extravagantly violated in order that the test be invalid. Robustness means that a parametric test can be used so long as the violation of the assumption is not too extreme. It is important to know where a parametric test is robust. (We can think of a parametric test that is robust as a four-wheel-drive car or a light-duty SUV. It can't go far off the road, but it can ride rougher roads than a parametric test that is less robust.)

The most important assumption that can be neglected for most parametric tests is the assumption of an interval level of measurement. So long as the data are scaled at an ordinal level with at least five levels, traditional parametric tests can be used reliably. (This is one reason why so many questionnaires allow five levels of response, such as "never," "rarely," "sometimes," "often," and "all the time.") Many parametric tests, especially the *t* test for two-group experiments, are robust to violations of normality. If the variables are distributed somewhat non-normally, the test is still good. The bad news is that the most common violation of normality, called a "heavy-tailed distribution" is also the one to which most parametric tests are most sensitive. Finally, there is an important assumption that cannot be violated. In general, parametric tests are not robust when different groups have different variances.

CRITICAL CAUTION

Most computer programs that do statistical tests also automatically check for violations of their assumptions. Any warnings issued should be taken seriously.

When the data violate assumptions of a test beyond the limits of its robustness, an alternative test must be used. Nonparametric tests tend to require fewer assumptions and those assumptions are different than those of the corresponding parametric tests. The traditional reason for using a parametric test when possible is that parametric tests are generally more powerful for the same *N*. A parametric test is more likely to tell us when an intervention has had an effect at lower cost.

KEY POINT

In business, statistical power often translates into cost savings. A more powerful test means a greater ability to detect the effectiveness of an intervention for the same cost. Power means more bang for the buck.

SURVIVAL STRATEGIES

The world is changing. Faster and cheaper computers mean that complex calculations cost hundreds of times less than they did only a few years ago. There is one kind of nonparametric test, called a *permutation test*, that is not generally less powerful than a parametric test. Permutation tests may be more or less powerful than parametric tests, depending on the data. They require very few assumptions. The reason they were not used very often in the 20th century is that they require an enormous amount of computation. However, there are now commercially available computer programs that do permutation tests on personal computers. The 21st century is here. Knowing about permutation tests can make you a hero at your firm.

The key to parametric tests is their simplicity. They test for changes in a specific parameter, such as the mean, and require that other aspects of the data be well-behaved. When what we want to know can be detected by a change in the mean of some variable, and the data are good, parametric tests are the standard. Sometimes, however, we are interested in changes to other features of the data. A nonparametric test may be available that tests for exactly the change we care about. If not, many nonparametric tests test for *any* change in the distribution between the groups.

Chi-squared tests for count data

There is one very common situation where nonparametric tests are needed and there is a favored nonparametric test, called the χ^2 test (pronounced "chi-squared") that handles it. The χ^2 test will be discussed in detail in Chapter 14 "Nonparametric Studies." For now, it is a good illustration of the value of nonparametric tests.

Up until now, we have been concerned with numerical variables, where the mean is a meaningful statistic. But what about our sheep? The different

Table 9-2 Cross-tabulation for χ^2 test: Sheep by color and type of wool.

	White	Black	Total
Heavy wool	42	6	48
Fine wool	76	14	90
Total	118	20	138

breeds and colors of sheep are not ordered. There is no "average breed" or even an "average color." If we are interested in seeing if an intervention affects the distribution of a nominal variable, we definitely need a non-parametric test. The χ^2 test is designed exactly for such a situation.

Suppose we wanted to see if the breed of sheep affected the color. If so, the proportion of black sheep for one breed would be *significantly* different from another. There are a number of different ways to define a proportion as significantly different. The χ^2 test defines a difference in proportions between groups in terms of the overall ratio of the dependent variable (in this case, color) for the entire population (in this case, the flock).

The χ^2 test is based upon a cross-tabulation of the two variables of interest. Table 9-2 shows a cross-tabulation for a flock of two types of wool sheep. (One of the requirements of the χ^2 test is that there be at least five subjects in each cell in the table, so we need to use a bigger flock than we did earlier.) The bottom row of the table contains the *marginal totals* for color. The right-most column contains the marginal totals for breed. The χ^2 test uses a statistic, called the χ^2 statistic, which compares each number in each cell to what that number would have been had the subpopulations defined by each value of each variable had the same proportions of the other variable as do the marginals. If the proportion of black and white sheep differs in any way between the two types of merinos, then the proportion for one breed will be higher than the proportion for the whole flock and the proportion for the other breed will be lower. In such a case, the χ^2 statistic rises.

The distribution of the χ^2 statistic for different degrees of freedom (that is, values of N) is known. As with the normal distribution and the t distribution, this means that whenever we can calculate the value of the χ^2 statistic, we can translate this into the probability of Type I error. The χ^2 test is called distribution-free not because there is no distribution for the χ^2 statistic, but because we do not need to know the shape of the theoretical distribution of color or breed in order to use the χ^2 test.

Statistical Significance In Business

As we suggested above, because the idea of statistical significance was developed in a scientific context, we have to be careful in applying it to business decisions. We can still make use of the concepts of Type I and Type II error, but we need to think about them in the context of business with the goal of maximizing return on investment.

Both the practices and the goals of science that guided Fisher's conservatism are different from the practices and goals of a business. In terms of practices, Fisher assumed that the current theory—expressed as the null hypothesis—had been established over the long term, through scientific inquiry. In a business, we may often think that the current theory or business model is the result of blind conservatism and unwillingness to look at facts. Or there may be no way to establish which theory is currently accepted; we may be simply looking at two alternate theories. For example, in our case study, one group of executives within the company may hold that sales training is not cost-effective, and another group holds that it is. Science moves forward by consensus. In business, consensus is only valuable when it serves higher goals.

In science, a standard α-level is set to limit Type II error over time and across experiments in order to keep the overall body of scientific knowledge stable as scientific progress causes it to change. These sorts of global, long-term concerns are not relevant to business decision support, where the goal is to make a decision that maximizes return on investment. It is important to note that the projected return on investment is not a purely financial figure. We must consider risks and ancillary effects that may improve or damage the long-term viability of the business. For example, doubling our profits for a year may not be worth the trouble if it forces the company to bankruptcy in 13 months. Therefore, we should do the following in planning to use the results of an experiment or study in support of a business decision.

- Turn our business questions into statistical questions for the experiment, and define the expected results in terms of hypotheses.
- Determine if one hypothesis has a genuine basis for being favored over the other(s), and should therefore be considered the null hypothesis. A null hypothesis must involve lower cost and predict no effect of the intervention.
- Define the business results that would arise from the predicted values based on different business models or hypotheses, using return on investment measures if possible.

- Define what non-statistical considerations, including corporate values and qualitative risk factors, should go into the decision-making process.
- Define the range of statistical results that would favor each hypothesis, and the range of results would be statistically non-significant, such that the results of this experiment or study should not influence the decision either way

This discussion is most applicable to studies that compare two options, and seek to determine which one is best for the business. However, in business, we almost always have more than two options. In these cases, the work of determining the direction that the results of our study truly support is challenging, and can be disputed easily. In such cases, it might be better to take an approach we suggested in Chapter 4 "What Is a Statistical Study?" and use the study to gather information and understand the issues, rather than to come to a particular decision. In this case, the study would be designed to help us understand:

- What are the most important questions to ask?
- What are the factors that appear to have significant correlation to desirable results?
- What further studies should be done?
- What conclusions or courses of action, if any, are supported through this initial study?

USING HYPOTHESES FOR BUSINESS DECISIONS

If we do a power analysis, we can determine the likelihood of both Type I and Type II error. If we can determine the costs and benefits of each type of error, we can use planning techniques to determine, first, whether it is worth the cost of the experiment to do the statistical testing at all, and, second, how we should act on our business decision depending on the results of the statistical analysis.

Designing the experiment to deliver maximal information value as an experiment

In this subsection, we will assume that we want to maximize the effectiveness of the statistical analysis. To do so, we design the experiment to maximize the variance due to the intervention, and minimize the amount of variance due to everything else.

Let us return to our example. We are designing a study to support a business decision regarding whether or not to institute a particular training program for our salespeople. We know the cost of setting up the program, as well as the cost per salesperson of providing the training. We know how much each additional dollar in sales means in increased profits to the company. A simple experiment would be to take the next N new hires to the sales force, divide them into two groups, train all the new salespersons in one group, and measure the difference in average sales between the two groups over a period of time. If the trained salespersons can collectively earn more for the company than the cost of the training, then the right decision is to start the training program.

But how big should our experiment be? The more new hires we train, the more expensive our experiment will be. We need to make N large enough so that it gives us an accurate enough measure of the effect of training, so that we can base our decision on it. We need to minimize both Type I and Type II error. We don't want to spend money on an experiment that gives us the wrong advice. The cost of a Type I error (where we mistakenly initiate the full training program, despite the fact that it is not worthwhile) is the cost of the experiment, plus the cost of the training program. The cost of a Type II error (where we mistakenly choose not to have a training program, despite the fact that it would increase our profits) is the cost of the experiment, plus the loss of the net benefit of having the training program.

The Type II error depends upon the variability of the amount of sales made by our untrained salespersons. We can estimate this variability by measuring the variability found amongst our current salespersons. (Our current salespersons are a sample of the population of all salespersons we might have, in the past or in the future.)

In order to do a power analysis to determine the likelihood of Type II error, we will need to know how big an increase in sales the training is going to produce. Unfortunately, this is the thing we hope to learn from our experiment. How do we make a determination of the size of our experimental effect before we do the experiment? There are two approaches. We could look at the experience of similar firms who have used similar training programs and guesstimate the smallest effect the training might have. We would then calculate the N required to detect that small of an improvement.

A better approach would be to pretend we have already done the experiment and use our planning tools to see just how big a difference a training program would have to make in order to have it pay for itself. This is an application of the return on investment (ROI) model to the design of our experiment. When looking at the value of training, ROI is called ROTI (return on training investment). If the training had a beneficial effect too small to pay for itself, we wouldn't want the new training program in any case, so that is the smallest effect we need to be able to detect. We do a power analysis and find out just how large an N we need in order to detect that much improvement due to training.

If the N required is so large that we cannot even afford the experiment, then we will need to make our decision without the benefit of statistics. If we can afford the experiment, of course, we should do it. The results will give us an estimate of how much sales will increase per salesperson over time. We don't even need to translate this estimate into a probability that the training program has had a significant effect. We already know that there is some probability of both Type I and Type II error, but we have lowered both of these probabilities to the point where we can afford to rely on our estimate. We can now use the results of our experiment directly with our planning technique in order to decide whether or not to institute the new training program.

Designing the experiment for optimal business results

If this experiment is going to guide long-term business results resulting in a very large change to net revenue, then a well-designed experiment—aimed at the greatest chance of providing useful information to guide our decision—may be the single most important consideration in our planning of the experiment. However, realistic business considerations are likely to intervene, requiring a change away from the best possible experimental design, due either to constraints or to competing objectives.

Let us look first at the issue of constraints. Here are some examples of constraints that might cause us to alter the experimental design.

- If we have a small, or medium-sized firm, our total number of sales people may be too small to allow for a study that will produce statistically significant figures.

- If our financial situation is desperate, not training some salespeople might drive us out of business by the time we see that training would work. If operating as we are now is too costly, it might be better to take the risk based on immediate professional judgment, rather than incurring the opportunity cost of the study, including the lack of improvement of sales figures for the control group.

- If there is a hiring freeze, we may have to choose to create an experimental group and a control group from our current sales staff, rather than from a group of new hires. From an experimental perspective, this is a disadvantage, because the population is more diverse in age and experience, introducing many new variables that might end up being the cause of error variation in the experimental result.

In short, a medium-sized or large company in good financial condition is more in a condition to try an experiment than is a smaller company, or one with fewer resources.

The study designed in this chapter so far has been focused on asking a rather idealized question: Will this one method of training increase gross sales for new hires enough to justify the investment in the training? Business decisions are rarely so simple. In the business world, we are more likely to have many options. For example, we may want to compare two training programs or we may want to choose to train certain employees, and to ask which ones would be most likely to benefit from training.

Fortunately, statistics has many tools that can help with these more complicated decisions, including both the design of quasi-experiments and the application of particular statistical techniques. Here are some examples of useful design of quasi-experiments and statistical analysis.

- *Creating a pair study*. Consider the case mentioned above, where we want to test the value of training for current employees. Rather than a random division of our sales staff into a control group and an experimental group, we could design an experiment called a pair study. In a *pair study,* we organize our random sample of subjects into pairs, just like in the buddy system. Each pair consists of two subjects who are as similar as possible in all criteria we consider relevant: age, years of experience in sales, years of experience in our company, gender, and so forth. Then, for each pair, we flip a coin and assign one member to the experimental group and the other to the control group. The control subjects do not receive the training, and we compare their results to their buddies who do.

- *Using additional surveys and polls*. It will be useful to gather as much information as possible about both our experimental group and our control group before, at the beginning of, during, and at the end of our study as possible. This information will support us in making the best decision among many choices, rather than a simple binary decision: provide training, yes or no? When we begin an experiment, we don't know what the results might be. That is obvious, but it is worth restating, because we easily fall into the trap of picturing two normal curves next to each other, and some result that falls somewhere on them. But consider this possibility. We conduct the experiment, and the outcome for the experimental group is bimodal. One mode centers around the old mean. For this subgroup, the training had little or no effect. The other mode is well above our goal based on desired ROTI, indicating that, for this subgroup, the training was highly successful. If that's all we know, we're dead in the water. But if we know a lot more about our experimental group, we may be able to define the common factor(s) that make them a unique subgroup. We may find they were enthusiastic about the training, either before it started, or when they came back. We may find that they had one particular trainer, or work under one particular sales manager. We may find that they all did one particular step the trainer recommended, where the other group did not. Whatever the case might be, the more data we have, the more likely we are to be able to recommend a particular training plan—who should be trained, in what ways, with what kind of follow-through—to maximize return on training investment.

- *Creating more complex experiments.* If we want to evaluate two or more different training programs, we can create a control group plus several experimental groups. Or, if we know we are going to do some training, but don't know which, we could create a quasi-experiment with no control group, and just groups who receive different kinds of training.

When business objectives conflict with good experimental design

Consider the following scenario. Due to the seasonal nature of our business, we have determined that we need a full year of gross sales figures from the experimental group and the control group to determine ROTI before we decide whether to adopt a ten-year training program. We run the study. In the first six months, we see a significant difference between the trained group and the untrained group. However, in the second six months, the two are much closer to one another.

One explanation is that the benefit of training is temporary. But there is another possibility. Suppose some friendly salespeople, or a smart sales manager, said to the control group, "Hey, those guys who got the training are doing a lot better than you in the control group. Let's set up a learn-at-lunch seminar, and they'll teach you what they're doing right. Your sales will improve." Well, that's a brilliant business idea—in-house cross training. And we've just ruined the experiment by contaminating the control group with what the experimental group learned in training. The result—the company makes an extra $10 million during the year of the experiment, but loses the benefit of the experimental study.

Sometimes, we might make such a choice intentionally. Every once in a while, a new drug being tested is found so effective that, before final Food and Drug Administration (FDA) approval, the FDA allows for treatment of individuals in the experiment—experimental subjects, control group members, or both, to receive the medication on a compassion basis. This requires violating the double-blind protocol, but it is done in the interest of compassion. Similarly, we need to decide when canceling or modifying a study makes good business sense. However, it is better that we know and manage what is going on, than that we find out after the fact that someone has taken an action—however good for the business—that reduces the integrity or value of our study.

WHAT'S SIGNIFICANT IN SCIENCE MAY NOT BE IMPORTANT TO BUSINESS, AND VICE VERSA

Some books of business statistics recommend the use of an α-level of .10, double the scientific level of .05. The idea is that being 90% sure is enough certainty to make a business decision. Certainly, an α-level of .05 is both arbitrary and was invented for a scientific purpose that has no meaning in business. As such, it is hard to justify keeping the scientific standard for business. On the other hand, an α-level of .10 is equally arbitrary. Furthermore, there is no reason for there to *be* a single standard for all business decisions. Science is a communal enterprise and needs a community standard. Business requires individual business decisions, each made on their own merits. As we saw in the preceding examples, a combination of planning methods and power analysis will help us make our business decision more effectively than a simple determination of statistical significance.

Of course, sophisticated decision analysis requires expertise. The use of the .10 standard can be appropriate if we are careful about the costs of Type II error and understand what an α-level of .10 actually means.

THE NEED FOR THEORIES DOESN'T APPLY TO BUSINESS

One element of the Fisherian system clearly has no place in most business decisions. There is no need to tie our hypotheses to competing theories of the world. In business, we are not seeking greater knowledge, but greater profits. Instead of linking the various hypotheses to theories, we need to link them to alternative actions in our business decisions. As shown in our example above, one way of doing this is to link the probability of Type I and Type II errors to their costs and benefits. But different types of decisions will make different uses of the results of our statistical analyses. The important thing to realize is that statistical significance requires a concept of hypothesis. We need to make effective use of statistical hypotheses in our business decisions in order to make effective use of inferential statistics.

THE NEED FOR EXPERIMENTS

Inferential statistics was invented around the idea of evaluating controlled experiments. It is entirely possible to use inferential statistics outside of experimental settings, or even when there is no intervention. The example of using the χ^2 test to see if breed affects the color of sheep is a good example. There is no intervention that magically transforms a sheep from one breed to another. If the χ^2 test for the sheep is significant, we know that it is unlikely that the different proportions of black sheep found among the different breeds is due to something other than the difference in breed.

What we lose when we don't do an experiment are the benefits of experimental control. If we don't do the intervention based on random assignment to groups, the difference we detect with our inferential statistics may be due to something entirely outside our control that just happened to correlate with our dependent measure. For example, suppose that, instead of training half of our newly hired salespersons as in our earlier example, we just compared N of our new hires to N new graduates of an independent training institute. If the graduates of the training institute outperformed our untrained hires, it might be because of the training, but it might be because better salespersons tend to go to the training program rather than answer our employment ads. Or it might be something else entirely.

However, the statistically significant result would still apply to the difference between graduates of the training institute and our hires. We might want to seriously consider recruiting from the graduating class of the institute.

Almost any descriptive statistic can be converted into an inferential statistic, or, at least, an equivalent inferential statistic can be found. Inferential statistics, in the last analysis, are just descriptive statistics with a limit placed on Type I error. The benefits of statistical significance do not transfer when we simply measure something with an inferential, rather than a descriptive, statistic. It is the entire structure of experimentation, or, at the very least, quasi-experimentation, that makes inferential statistics valuable in assuring that our Type I and Type II error levels are accurate and mean something useful. It is a very bad practice to simply calculate the P-value for a statistic that is just a measure of some event in our business. People associate P-values with a certification that the results are reliable indicators of a cause and effect relation. *P*-values are only truly meaningful in the context of a properly controlled intervention. Using them elsewhere can be deceptive.

The most common misuse of *P*-values is with the correlation coefficient. Correlations can be taken between any two variables at any time. It is common (and useful) to take correlations outside of an experimental setting. There is also a way to derive a *P*-value from any correlation. Under very narrow and rarely realized conditions, the *P*-value for a correlation coefficient tells us that the correlation is not due to chance alone. Outside of an experimental setting, the best solution is to report just the correlation coefficient and leave the statistical significance alone.

CRITICAL CAUTION

Correlation is Not Causality

Within the context of a well-designed experiment, we can assert that correlation between the intervention and a difference between a value for a particular variable in the experimental group from the value found in the control group is an indicator that supports the idea that the intervention caused the difference in the value of the variable. However, this assertion depends on many aspects of a well-designed experiment. For example, experimental design ensures that the intervention preceded the measurement of the variable, that no other event might have caused both the intervention and the difference in the value of the variable, and that many other possible sources of variance were minimized.

None of those is true for statistics gathered from real-world (non-experimental) data, from surveys, or from polls. As a result, if we find a strong statistical correlation between variables *A* and *B* in data gathered, but not from an experiment, all we can do is assert a correlation. We may not be able to determine whether *A* always preceded *B*, or vice versa. We can't know if there was some preceding factor *C*, that we never thought to measure, which caused both *A* and *B*.

The lesson: Correlations outside experimental contexts do not necessarily point toward causality.

On the other hand, there is good news for business. Sometimes, it is enough to establish correlation. In certain business situations, we can make use of a correlation for business purposes, even if we don't know its cause. For example, if we find that fans of a certain pop singer buy our brand of toothpaste, we can advertise at her concerts, even if we never find the interview in *People Magazine* where she said that our toothpaste was the key to her fame and fortune.

Statistical significance always looks impressive, but it is not always meaningful.

Quiz

1. The "alpha-level" for accepting a new theory over the old theory is...
    (a)   1 in 5
    (b)   1 in 20
    (c)   .20
    (d)   .50

2. Fisher's strategy of statistical inference protects us from...
    (a)   Falsely abandoning the current theory
    (b)   Falsely abandoning the new theory
    (c)   Falsely accepting the new theory
    (d)   All of the above

3. What type of research design provides Fisher's guarantee of conservatism?
    (a)   Correlation
    (b)   Ratio
    (c)   Experiment
    (d)   Proportion

4. For a minimum group test, we need...
    (a)   Control group
    (b)   Experimental group
    (c)   Both (a) and (b)
    (d)   Neither (a) nor (b)

5. The variable that is measured is the _____ variable.
    (a)   Dependent
    (b)   Independent
    (c)   Control
    (d)   Experimental

6. The variable that defines the groups in a group test experiment is the _____ variable.
    (a)   Dependent
    (b)   Independent
    (c)   Control
    (d)   Outcome

7. If each subject has an equal chance of being in the control group or the experimental group, they have been _____ to groups.
    (a)   Hypothesized
    (b)   Null hypothesized
    (c)   Randomly sampled
    (d)   Randomly assigned

8. A Type I error occurs when we ...
    (a)   Reject the null hypothesis when it is true
    (b)   Reject the null hypothesis when it is false
    (c)   Accept the null hypothesis when it is true
    (d)   Accept the null hypothesis when it is false

9. Which of the following is an example of a quasi-experimental design?
    (a)   Pair studies
    (b)   Polls
    (c)   Surveys
    (d)   All of the above

10. In a business context, a correlation cannot tell us ...
    (a)   The relationship between two variables
    (b)   Statistical significance
    (c)   Causality
    (d)   Error

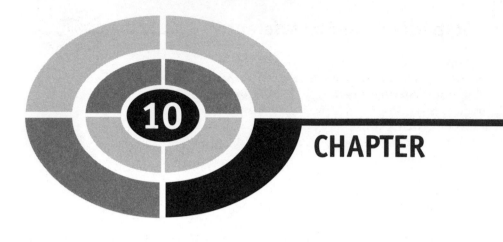

10 CHAPTER

Reporting the Results

This chapter shows you how to take the last step in creating a statistical report—the preparation and delivery of the written report, oral presentation, or presentation in any other medium appropriate to the business goals of the statistical report that presents the results of the study or research you have performed.

Three Contexts for Decision Support

Our statistical research will, in most cases, be used to support decisions in one of these three situations:

- *Input to a decision.* We will deliver our report internally to a group holding a meeting or series of meetings, or to an executive making a decision. Our report will be one input for the decision process.

- *Support for a decision when it is announced.* Once a decision is made, if it was supported by statistics, our company is likely to want a version of the statistical report presented as a justification for the decision. This may be an internal document, or it may be intended for stockholders, employees, vendors, customers, or the general public.
- *Advertising.* Advertising is a special case of decision support where the decision is being made by the company's customer, not by anyone inside the company. The information from our study or research can be used to guide the customer in making an informed choice of whether or not to purchase our products or services, and which ones to acquire for what uses.

First we will offer some guidelines for creating good reports and presentations in general, and then we will look at each of these three specific situations, and how to create a report or presentation for each.

Good Reports and Presentations

A good report has the qualities of a good technical specification, plus other qualities that make it appropriate for our audience. Table 10-1 shows a modified and simplified version of the qualities of a good technical specification taken from the IEEE (Institute of Electrical and Electronic Engineers) specification, as presented in Karl Weigers' *Creating a Software Engineering Culture*.

We should prepare a complete set of technical results in tabular format, with the most important charts and graphs as well, meeting all of the requirements in Table 10-1 before we begin to prepare the reports we will deliver to our audiences. If we prepare this set of master documents well, then generating high-quality reports in various formats will be much easier.

GOOD REPORTS

Once our master documents are prepared, we can turn our attention to our audience. For the design of a good report, see Chapter 4 "What Is a Statistical Study?" Some people find it easiest to write the summary first, because it works well to begin with the audience in mind. Others—especially those trained in research methodology—will prefer to begin with the data and statistics, and work towards the conclusion. Either approach is fine. However, as we edit, we should work in the opposite direction from the way we wrote. If we began with the conclusion, we should validate our results by

Table 10-1   Characteristics of a good report.

Characteristic	Description	How we ensure this quality
Complete	Nothing is missing: The entire topic is covered to meet the reader's needs	Define the goals of the study for all audiences. Create an outline of the study that meets these goals. Review a draft of the study for completeness. Consider what appendices and extras would meet reader goals.
Consistent	The report contains no internal contradictions. Where it contradicts existing ideas or published works, it identifies that clearly.	Create a data dictionary, variables list, and glossary early. Maintain them throughout, and check all tables and graphs against them to ensure consistency, as well as the final report itself. Audience: Know your audience and the language and ideas they already accept and understand. Have the report edited by someone who can understand the topic. Ensure two pairs of eyes see every table, figure, and paragraph. Prepare lists of tables and figures as you create the report.
Correct	The book accurately reports the process and results of the research, experiments, and analysis.	Follow appropriate protocols and standards for research and analysis. Define your fact checking needs, and get the work done early.
Feasible	The study or research is a well-defined project delivered with quality, on time, and within budget. The report is available in both summary and full form.	Use a level of project management appropriate to the size and complexity of the project. Get expert help for statistical and experimental design. Work with the report recipients early to define the forms of the report that will be useful for them. Prepare a mock-up of the report and get the format approved.
Modifiable	The report can be changed or emended, if necessary.	If the work is to be done only once, simply maintain version control of the document and retain an editable electronic copy. If similar studies might be conducted in the future, prepare a research plan and templates so that future editions can easily be prepared in the same format.

Table 10-1   Characteristics of a good report (*Continued*).

Characteristic	Description	How we ensure this quality
Necessary	Meets a need or needs of the reader, providing appropriate decision support.	For the entire report, this is established at the beginning of the study. Be sure to link each section, figure, and table of the report to the purposes of the decision meeting.
Prioritized	Results are ranked by significance.	Evaluate significance of each finding in relation to the decision being supported. Note that reporting of results that show the study cannot support one decision over another are significant to the business.
Testable	Experiments, quasi-experiments, studies, and analyses can be reviewed or repeated.	All activities of the study have a full set of notes and an audit trail. This information is listed in the appendix and available in an appropriate archive.
Traceable	Each element is uniquely identified so that its origin and purpose can be traced to ensure that it is necessary, appropriate, and accurate.	We may not include full notes in the presentation copy of our report, but we retain a master copy that allows us to trace each figure, table, and assertion in the report back to the analysis data that support it. All of our planning, research, experiments, and analysis include a thorough audit trail, set of research notes, and copies of questionnaires, design documents, and data at all stages of our work.
Unambiguous	Each element has only one possible interpretation.	Have an editor who did not participate in the work read and edit the report, identifying any uncertain or unclear ideas. Make sure that the report is in language understandable to the audience. Be sure that any terms that have different meanings in different fields present in the report are used clearly. For example, the term "significant" has different meanings in decision support and in statistics.

walking through the logic from our data to that conclusion. If we began with our data, we should read the report from the executive perspective, making sure that the results are presented in a way that supports the decision, or justifies it, or supports advertising, in accordance with our objectives.

STUDY REVIEW

Input for the Next Step

Remember that your reports and presentations are your deliverables, but, for business purposes, what matters is that they are inputs for the next step—the decision process, publication, or advertisement. Therefore, quality is measured primarily by usability. If your work is questioned, you should be able to defend it, but a review of your process is not a normal part of the next steps. Business tends to move forward, not to look back.

GOOD PRESENTATIONS

Presentations, whether given at meetings or telemeetings, or presented in automated formats—anything from web sites to TV commercials—are different from reports in one major way. Our audience can read a report at their own pace, and re-read it if necessary. However, in a presentation, the information comes by only once. As a result, we need to be very careful to design in appropriate emphasis of key points, and repetition where it is helpful. Of course, if the audience of the presentation receives the executive summary or report before or during the presentation, this will help them understand the material.

In preparing to present, we face the same challenge. In writing a report, every correction accumulates, improving the document. In preparing a presentation, we must rehearse many times, and, even so, we may make mistakes on the big day that we have corrected over and over.

SURVIVAL STRATEGIES

Practice, Practice, Practice

Find the time to rehearse your presentation. Repetition, plus observing yourself, is irreplaceable. A good guideline is that you should rehearse a presentation in full three to six times before giving it. Even professional speakers follow this rule. This applies to all oral presentations, both in business and in the classroom.

Here are some tips for organizing a presentation:

- *Create an outline.* Include an introduction, the body, and a conclusion.
- *Know your role.* Are you simply presenting the report? Or are you guiding the discussion? Who speaks before you, and after?
- *Be ready to ask questions.* Asking questions and doing other things that make the presentation interactive helps ensure your audience understands.
- *Be ready to answer questions.* Prepare a document with notes, questions, and answers for your own reference. When people ask questions, feel free to refer to your notes.
- *Prepare a presentation of the appropriate length.* If you are creating a slide show, plan for two slides per minute of the presentation, but less when you are explaining figures and tables at length. Rehearse and measure the time. Plan the presentation to end a few minutes early, and have some optional material ready in case there are no questions.

Here are some tips for getting ready to give a presentation:

- *Know your material.* Rehearse to the point where you only need to look at your notes occasionally. Practice—or simplify—difficult words and concepts.
- *Go slowly.* You can stop and breathe after each slide, and people will not think you are going to slow. An intentional pause, like an intentional gesture, gives rhythm and sets the pace. Actually, you will seem slower if you keep talking longer without focus, or if you fill pauses with ums and ahs.
- *Do what is comfortable.* If it is easier, stay behind a podium. If you are comfortable moving around, making gestures, and so forth, do so. Anything you do with clear, focused intention works. But shifting or rocking back and forth unconsciously is a distraction.
- *Stand with one foot in front of the other.* This will keep you from rocking back and forth.
- *Get comfortable with your presentation tools.* Whether you use a data projector, a flip chart, or anything else, learn to handle these tools effectively, so that you can keep your focus on your audience.
- *Get help from experienced presenters.* Colleagues, or a local chapter of ToastMasters or the National Speaker's Association are great resources, and very generous with help and guidance.
- *Know that your audience supports you.* Know that everyone wants you to succeed—as much for their own sake as for yours. No one enjoys being at a poor-quality presentation, or seeing someone make an

embarrassing mistake. If you handle any mistakes professionally, moving past them, others will overlook the errors with you.

Reports and Presentations Before the Decision

When providing a statistical report as input to a decision process, our focus should be on presenting the question and presenting whatever guidance the results of the statistical study or research offers in making the decision. The clearer the question is to everyone, the easier it is to present the study in support of the decision.

It is sometimes difficult to present the idea that the study returned no significant results. Work carefully to explain how a lack of statistical significance is a valid result of a study. Also explain what it implies for the decision to be made.

As you plan your reports and presentations, be clear what you are trying to accomplish in each section of the report or presentation. Here are the different goals you may have:

- Define the decision to be made, including two or more options, one or more of which will become the course of action.
- Define the questions being asked.
- Explain technical or statistical issues to a non-technical audience.
- Explain the course of action recommended by the statistics.
- Explain how the statistics returned no statistically significant results, and what implication this has for the business decision.
- Make recommendations for further study, if appropriate.

CRITICAL CAUTION

Be Wary of Political Pressure
Some people—perhaps your superiors—may enter the decision meeting already biased or decided about the way the decision should go. You may be asked to alter your conclusions or recommendations. Or you may be asked to minimize the significance of your results. On the one hand, you should be willing to present findings in appropriate and inoffensive ways. On the other hand, it is essential that you not compromise the truth.

Reports and Presentations After the Decision

If a decision is made, and a new course of action is being launched, then there will be some sort of announcement—internal or public—of the decision. This is likely to include some explanation of the reasons for the decision, including the statistical results that support it. This could be a simple sentence, or it could be a summary of the statistical report, including some tables or charts.

If the document is to be a formal report—such as a corporate annual report—or a page on the corporate web site, then you will need to coordinate the format of your report with the people responsible for the design of these documents. If you are preparing a report or presentation for the public, be sure to coordinate your work with the publicity department or other official spokespeople for your company.

You may be asked to prepare or present multiple reports or presentations for different audiences. In that case, be sure you understand each audience, their needs, and what information the company desires each audience to receive, with what focus. At worst, this is a matter of putting different spins on the same information. At best, it is a positive effort to engage the support and understanding of groups with different interests and concerns.

If you are asked to give a presentation in an unfamiliar forum—such as a presentation to a large, public audience—request whatever support you need to prepare.

CRITICAL CAUTION

Statistics Gathered After the Fact

Sometimes, you may be asked to prepare a statistical report that supports a decision after it has already been made. Or you may be asked to alter results so that they support a decision. In the first case, this risks the post-hoc fallacy. In the second case, there is a risk of falsifying results. If such a situation arises, it is best to provide only those statistics that support the decision, if any. If it is possible to gather statistics that genuinely support the decision without falling into the post hoc fallacy, that is also all right. It is within a company's right not to publish data and study results it owns that goes against a business decision it has made, but it is inappropriate, and, in the long run, disadvantageous, for a company to present results that are not statistically supported, or worse, fallacious.

Advertisements and Sales Tools Using Statistics

There are many ways that statistical studies can be used in the design of advertising campaigns and in the content of advertisements. Marketing studies can be used to define the audiences, themes, and content that are likely to convince people to buy our products and services. In addition, data about the benefits of our products and services can be gathered through a study, analyzed with statistical techniques, and presented to customers.

In almost all cases, advertising will simplify—perhaps dangerously oversimplify—the results of statistical studies. As we provide material to be used in advertising, we should try to do the following:

- *Write simple, clear statements that are appropriate to your company's potential customers.* This way, advertisers will be less likely to need to make changes—and perhaps introduce error—in the presentation of statistical results as advertising content.
- *Make a list of statements that cannot be supported by the statistics.* Think of things that the advertising writing team is likely to want to say, and let them know what is, and is not, supported.
- *If possible, join in some of the brainstorming sessions.*
- *If possible, ask to review the advertisements before they go final.* This may give you a chance to correct misstatements before they are distributed. Make sure that you know whether you have the authority to correct errors, or only the right to advise.

This work will be easier if you familiarize yourself with the types of advertising and campaigns that your company engages in. In some cases, you may be asked to offer more sophisticated studies in support of customer decisions. For example, if your company's clients are businesses who purchase large contracts or quantities of products from your company, you may be able to perform statistical analyses in support of proposals that show the value of your company's services for a particular contract.

CRITICAL CAUTION

Advertising Knows How to Misuse Statistics
Many techniques for exaggerating and distorting statistics have been used in advertising for over half a century. For a good list, read *How to Lie With Statistics*,

which was written in 1954. Many practices which are unacceptable to statisticians are common in the field of advertising. This leads to tension, to say the least. It can be challenging to present a case for the business value of honesty in this environment.

Quiz

1. Statistical research will be used to support which of the following decisions?
   (a) A decision before it is made
   (b) Support for a decision when it is announced
   (c) Advertising
   (d) All of the above

2. By linking each section, figure, and table of the report to the purposes of the decision, we establish the _____ of the report.
   (a) Feasibility
   (b) Necessity
   (c) Testability
   (d) Consistency

3. By determining that the report contains no internal contradictions, we establish the _____ of the report
   (a) Consistency
   (b) Necessity
   (c) Testability
   (d) Feasibility

4. As we edit a report, we should work...
   (a) In the same order in which we wrote the report
   (b) From the executive perspective
   (c) In the opposite order in which we wrote the report
   (d) From the data to the conclusion

5. When organizing a presentation, it is good to...
   (a) Create an outline
   (b) Be ready to ask questions
   (c) Be ready to answer questions
   (d) All of the above

6. A report presented before a decision is made should focus on...
   (a) Clarifying the question and how the results support the question
   (b) How the results can be used in advertising
   (c) Being as technically specific as possible
   (d) Statistics gathered after the fact

7. In almost all cases, advertising will _____ the results of statistical studies.
   (a) Falsify
   (b) Simplify
   (c) Exaggerate
   (d) Misuse

8. A good guideline for rehearsing presentations is to rehearse it _____ times before giving it.
   (a) Once
   (b) 2 to 3
   (c) 3 to 6
   (d) 10 or more

9. To avoid the problems associated with using statistical studies in advertising, we should...
   (a) Write clear statements that are appropriate to your customers
   (b) Make a list of statements that cannot be supported by the statistics
   (c) Ask to review advertising content before they are final
   (d) All of the above

10. A good report has the qualities of a good _____, plus other qualities that make it appropriate for our audience.
   (a) Technical specification
   (b) Advertising campaign
   (c) Decision
   (d) All of the above

Exam for Part Two

1. Which of the following is not a reason for conducting a statistical study in a business context?
   - (a) Providing information in support of answering questions that are likely to arise
   - (b) Providing general information about a situation in order to define what questions should be asked
   - (c) Providing support to answer specific business questions and make specific business decisions
   - (d) All of the above are good reasons for conducting a statistical study

2. Which of the following methods does not necessarily involve asking questions of people?
   - (a) Opinion poll
   - (b) People poll
   - (c) Survey
   - (d) All of the above involve asking questions of people

3. The process of determining which subjects will be in the control group and which will be in the experimental group is known as _____.
   (a) Assignment
   (b) Sampling
   (c) Polling
   (d) Encoding

4. Turning a vague business question into a good research question requires which of the following?
   (a) Defining the population
   (b) Defining the attributes of the population you want to measure
   (c) Defining the significant precision for each attribute you want to measure
   (d) All of the above are required

5. Self-descriptions are examples of...
   (a) Facts
   (b) Opinions
   (c) Unbiased data
   (d) Always accurate

6. Including numbers in a statistical report adds _____ to the presentation.
   (a) Precision
   (b) Clarity
   (c) Correctness
   (d) All of the above

7. What three components need to be carefully planned for data collection?
   (a) The source of the data
   (b) The methods that ensure reliable and valid data
   (c) The research design that ensures the suitability of our data for our planned analysis
   (d) All of the above

8. In general, obtaining _____ is preferable to obtaining _____.
   (a) Archived data; non-archived data
   (b) Non-archived data; archived data
   (c) Detailed data; summary statistics
   (d) Summary statistics; detailed data

9. The advantage of collecting data yourself is that...
   (a) It is less expensive than purchasing data from a vendor
   (b) It gives you complete control over the data collection process
   (c) Both (a) and (b)
   (d) Neither (a) nor (b)

10. Every time data are copied, moved, or translated from one form into another adds the potential for _____.
    (a) Calibration
    (b) Redundancy
    (c) Data validation
    (d) Error

11. What is probably the most reliable form of data entry?
    (a) Humans
    (b) Audio/Video tape
    (c) Electromechanical devices
    (d) Computers

12. Which of the following is *not* true of a pie chart?
    (a) Each individual unit that is included in the whole must be uniquely identifiable as belonging to one and only one of the parts
    (b) The sum of the measures can exceed 100%
    (c) There must be an identifiable whole whose parts need to be identified
    (d) All of the above

13. A histogram should be used when the data are on a _____ scale.
    (a) Nominal
    (b) Ratio
    (c) Ordinal
    (d) Interval

14. A _____ shows how being high or low on one numeric variable relates to being high or low on a second numeric variable.
    (a) Scatter plot
    (b) Histogram
    (c) Line graph
    (d) Area graph

15. If "Time" is a variable in your study, you should use a _____ to represent it in a graph.
    (a) Bar chart
    (b) Line graph
    (c) Area plot
    (d) Scatter plot

16. The *degrees of freedom* is equal to...
    (a) Standard error
    (b) Variance
    (c) Sample size
    (d) Population size

17. If the sample size is large enough, the distribution of the sample will resemble the _____.
    (a) Population distribution
    (b) Variance
    (c) Standard deviation
    (d) Correlation coefficient

18. Which statistic is *sufficient*, a *parameter*, and a *moment*?
    (a) Mode
    (b) Mean
    (c) Median
    (d) All of the above

19. The _____ give us a measure of the dispersion of the distribution
    (a) Variance
    (b) Standard deviation
    (c) Both (a) and (b)
    (d) Neither (a) nor (b)

20. The hallmark of inferential statistics is...
    (a) Statistical significance
    (b) Correlation
    (c) Capitalizing on chance
    (d) All of the above

21. A Type II error occurs when we...
    (a) Reject the null hypothesis when it is true
    (b) Reject the null hypothesis when it is false
    (c) Accept the null hypothesis when it is true
    (d) Accept the null hypothesis when it is false

22. The most important assumption that can be neglected for most parametric tests is...
    (a) The assumption of ordinal level of measurement
    (b) The assumption of ratio level of measurement
    (c) The assumption of interval level of measurement
    (d) The assumption of categorical level of measurement

23. To determine the likelihood of both Type I and Type II error, we conduct a _____ analysis.
    (a) Frequency
    (b) Power
    (c) Significance
    (d) Ratio

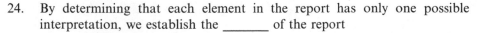
24. By determining that each element in the report has only one possible interpretation, we establish the _____ of the report
   (a) Unambiguousness
   (b) Modifiability
   (c) Traceability
   (d) Feasibility

25. Statistics gathered to support a decision that has already been made can introduce _____.
   (a) The risks of making the post-hoc fallacy
   (b) Falsification of the results
   (c) Both (a) and (b)
   (d) Neither (a) nor (b)

Statistical Inference: Basic Procedures

Part Three of *Business Statistics Demystified* covers the most common inferential procedures used in business statistics. An inferential statistical procedure is a method for generating a statistical measure along with a second statistical measure (usually a *P*-value) that tells us how confident we can be that the first measure accurately reflects the actual world. We have divided Part Three into four chapters. In Chapter 11 "Estimation" we introduce procedures for estimating population measures from samples. In Chapter 12 "Correlation and Regression," we look at procedures that provide measures that tell us about the relationship between variables, including the possibility that one variable measures a cause and another one measures an effect. In Chapter 13 "Group Differences," we explain analysis of variance (ANOVA) and designed experiments, offering procedures that measure whether or not an intervention on one group (or groups) has caused a difference between those group(s) and the control group(s). In Chapter 14 we will look at "Nonparametric Statistics," alternatives to the procedures in Chapters 11, 12, and 13, that may be less powerful, but aren't as picky about the quality of our data.

Part Three is a concise review of the most important tools for business statistics. It is a catalog. You will learn the details of each procedure from your professor and a textbook that covers these materials in depth. When it comes to running the procedure, any computerized tool will guide you. Our goal is to demystify the tools for you by providing a clear review of the most important issues, and to explain how they are used in business.

Note that throughout Part Three, N will be used to indicate the sample size. The population size is assumed to be infinite unless otherwise indicated. Also, because each chapter is short, we have not included quizzes. There is just one final exam for all of Part Three.

Estimation: Summarizing Data About One Variable

In this chapter, we address statistical estimation, the process of determining what we can say about the population based on a sample. In previous chapters, we discussed the importance of estimation, the general strategy for how to estimate, and the relationship between estimation and statistical inference more generally. In business, we may often want to know—or be able to claim—things about our population from just one sample. We may have contracts with our customers that promise a minimum proportion of some ingredient, a maximum amount of some impurity, a specific weight per unit, etc. Similar requirements may be imposed by various governmental or non-governmental regulatory authorities. Even if there is no contract or regulations, our advertising may promise that we meet some such standard. In these circumstances, it is useful to be able to assert our estimates with respect to some specific value with confidence in order to be able to guarantee our customers or the regulatory authorities that we have achieved some standard.

CRITICAL CAUTION

In Statistics, Estimates are About the Present, not the Future

In statistics, the term *estimation* refers to making statements about a population from a sample or samples. The population already exists, so the estimation is about the unknown present—the values of the population—not about the future. This will seem odd to many business analysts, including those who prepare budget estimates and time estimates, which are about the future. When statisticians talk about the future, they call it *forecasting*. When statisticians and business people get together in the same room, all get confused.

In this chapter, we will cover some of the more common estimation procedures used in business, step by step. Before proceeding with this chapter, it is a good idea to review the idea of estimation presented in Chapter 8 "Common Statistical Measures."

Basic Principles of Estimation

The idea behind estimation is that there is a value out in the world that is characteristic of a population and that by measuring some variable (called the *estimate*) from a sample of that population, we can get a good idea of what that population value is. The most common method of estimation is to take a sample of the population and calculate a statistical measure from the sample. The inferential method then allows us to calculate a second number that tells us how confident we can be about that statistical measure. Most often, this second number (or numbers) provides a measure of probability related to a numerical interval. It tells us how likely or unlikely it is that the actual population value differs from the sample value by that interval.

For example, we may want to list the weight of the contents of our snack-sized bags of potato chips as 1.5 ounces. Yet we don't weigh every bag, so how do we know? We weigh a sample of the bags, and collect those measures. The sample will have a mean weight, a variance, and a known distribution. And, of course, we know both the sample size and the population size. (Technically, the population size might be considered indefinitely large, since the label will be on the bag long after the estimation process is over.) With these figures, we will be able to say something like, "Our bags of potato chips weigh 1.515 ounces, ±1%, 99.73% of the time. Note that we've used three statistical figures: the sample mean (which estimates the population mean), the size of the confidence interval (which is a measure of the variability of the sample mean), and the type of the confidence interval (which relates the variability to the likely value of the true population mean). In this case, the confidence level is 3 sigma. And, since the lower end of our range is 1.505 ounces, we've left a small margin for error beyond that. If we can keep to such tight tolerances, we can ensure that we almost never sell underweight—a crime—while minimizing the extra cost of potatoes, oil, and salt due to over-stuffing our bags of chips.

POINT ESTIMATORS: ESTIMATING POPULATION VALUES FROM SAMPLE VALUES

Not all statistical measures of samples relate well to the corresponding measure for the population. For example, the minimum and the maximum of the sample do not relate to the minimum and the maximum of the population. (An infinite population may not even have a minimum or maximum.) An important reason why the mean is so commonly used in statistics is because it has a number of useful properties.

- The mean is what is known as a *consistent* estimator. No matter what the shape of the population distribution, as the sample size increases, the sample mean gets closer and closer to the population mean. (This is called the Law of Large Numbers. Of course, this is all a matter of probability. The sample mean does not get steadily closer. It is just that, the larger the sample, the more likely the sample mean is close to the population mean.)
- No matter what the shape of the population distribution, as the sample size increases, the sample mean is itself normally distributed. (This is called the Central Limit Theorem.) In other words, even if the population distribution is not close to a bell curve, the distribution of means of repeated samples of $N > 30$ will be normal. This is a very important property for making estimates, because it allows us to use the probability values of the normal distribution to express our confidence about how close the sample mean is to the population mean.
- The mean is what is known as an *unbiased* estimator. This means that, even for a given sample size, the mean of the sample distribution equals the population mean. In a sense, an unbiased estimator points at the true value exactly.
- The mean is what is known as a *relatively efficient* estimator for the normal distribution. Think of two different statistical measures, like

the mean and the median, which both measure the same characteristic of the population. If, for any sample size, one measure has a consistently smaller sample variance around the population value, then it is said to be more efficient than the other. If the population is normal, the sample mean is the most efficient estimator of the population mean.

- As we discussed earlier in Chapter 8 "Common Statistical Measures" the mean is a *sufficient* statistic. To say that a statistic is a sufficient estimator means that it uses all of the available information in the sample useful for estimating the population statistic. While the mean is a sufficient statistic, it is not always a sufficient estimator of the population mean.

So long as the population is normally distributed, the mean is a terrific measure for estimating the central tendency. Even if the population is non-normal, the mean is a very, very good measure. Whenever we can cast our business decision in terms answerable by finding out a central tendency for some population distribution, we can look to the mean as the best measure to use. This is why so many statistical procedures use the mean. On occasion, we may need to estimate some other characteristic of the population distribution. Under these circumstances, we should try to use a statistical measure that has as many of the above desirable properties as possible for doing the estimate.

FUN FACTS

A Baker's Dozen
Before it was possible to run production lines with close tolerances, the minimum was much more important than the mean when it came to delivery weight. The cost of a bit of extra for the customer was less more important than the cost of being caught selling underweight. The term "baker's dozen" for thirteen items comes from one solution to this practice. In England in centuries past, it was a serious crime for a baker to sell under weight. Government officials would come and check. But, with every roll hand-made, some would certainly be a bit too light. A baker could protect himself from criminal prosecution through the custom of selling a baker's dozen, thirteen for the price of twelve.

STANDARD ERRORS AND CONFIDENCE INTERVALS

When we estimate, we get at least two numbers. The first number is our best guess of the true population value. The second number (or numbers) is our best guess as to how far off our first number is likely to be, in other words, the error. When we report these two numbers, we need to add a third number that clarifies how the absolute size of the error relates to the likelihood of where the true population value lies. There are a number of ways of specifying this third number. As we discussed in Chapter 8, one way of characterizing the size of the error is by giving the standard deviation of the sampling distribution, the standard error. In terms of estimation, this is a somewhat awkward way to present the error. What the standard error tells us is that, should we repeat our study, our next value of the estimate is most likely to fall within the error bounds listed. In short, the standard error is not described in terms of what is being estimated.

An alternative for presenting error is a confidence interval. The type of confidence interval, expressed either as a percentage or in terms of sigma is our third number. The advantage to a confidence interval is that, the third number can be related to the population value. A ninety-five percent confidence interval for instance, is an interval surrounding the estimated value in which the true population value is 95% likely to fall. While true, this statement is slightly misleading. In ordinary English, when we say that a point is 95% likely to fall within some interval, we mean that the point could be in various places, but is 95% likely to be in the fixed region specified as the interval. We might express the expertise of an archer by saying that there is a 95% chance that her arrow, once fired, will land on the target. The location of the target is fixed. The arrow is yet to be fired. However, when we say that the population value is 95% likely to fall within the stated interval around the sample value, it is the population value that is fixed and the interval which would vary if we were to repeat our study.

It is sort of like the old joke about the guy who shoots an arrow at a fence and then paints a bull's eye around it. A confidence interval is like that guy having very poor eyesight. He hits the fence with the arrow and then feels around for the arrow and paints a circle around it. His eyesight is so poor that he can only be sure of surrounding the arrow with the circle 95% of the time.

Another analogy may help. Statistical estimation almost always involves fishing for an unmoving fish using a net. We toss the net, but then we are prevented from pulling it in. We can never verify that our net caught the fish, but we can express our confidence in our net-throwing accuracy by stating that, were we able to pull the net in, there would be a 95% (or however much) chance that the fish would be caught in our net. This percentage describes the proportion of our throws that would catch the lazy fish, not the likelihood that one throw would catch a moving fish.

THE z TEST

As shown in Table 11-1, the z test is a statistical procedure that allows the estimation of the population mean when the population variance is known.

Table 11-1   The z test.

Type of question answered	
Is the population mean significantly different from a specified value?	
Model or structure	
Independent variable	A single numerical variable whose mean value is of interest.
Dependent variable	None
Equation model	$z = \dfrac{\bar{X} - \mu}{\sigma_x / \sqrt{N}}$
Other structure	The P-value calculated from the z-score is the estimate of the probability that the sample mean would fall this far or further from the specified value, μ.
Corresponding nonparametric test	None
Required assumptions	
Minimum sample size	20
Level of measurement	Interval
Distributional assumptions	Normal, with known variance.

Single-Sample Inferences: Using Estimates to Make Inferences

It is only in rare cases that the population variance is known with such certainty that the z test can be used. When we have no independent source of information as to the variance of the population, we must use our best estimate of the population variance, the sample variance, instead. For example, we can't know the variance of the weight in the entire population of every bag of potato chips we sell, because we can't realistically weigh every bag. When we use the sample variance instead of the (unknown) population variance, we lose a degree of freedom. But the sample variance is also a consistent estimator of the population variance, so the quality of the estimate gets better with increasing sample size. We need to adjust our test to account for the sample size as a source of error in estimation. Recall from Fig. 8-4 that the t distribution changes with the sample size. As it turns out, the adjustment we need is just to use the t distribution for the appropriate degrees of freedom instead of the normal distribution used in the z test.

The single-sample t test is an excellent example of a common occurrence in statistics. The t distribution, which was originally invented to deal with the distribution of differences between normally distributed variables also turns out to be the distribution of a difference in means with a sample variance. The most common distributions have many different uses, because various kinds of estimates turn out to be distributed in that shape. On a practical level, the t test can be used easily because all of the input numbers are drawn from a single sample, like our sample of bags of potato chips.

HYPOTHESIS TESTING WITH THE t DISTRIBUTION

As shown in Table 11-2, the one-sample t test is a statistical procedure that allows the estimation of the population mean when the population variance is unknown and also must be estimated.

COMPARING PAIRS

As shown in Table 11-3, the paired t test is a statistical procedure that allows the determination of whether an intervention on individual members of multiple pairs of subjects has had a significant effect by estimating the population mean for the value of the differences.

From a design perspective, in terms of the type of question answered, the paired t test is really a group test, in that it can be used to measure the effects of an experimental intervention. We include it here, rather than in Chapter 13 "Group Differences," because in terms of the statistical calculations, the difference measure, D, is assumed to be calculated from a single sample of differences.

Table 11-2   The one-sample t test.

Type of question answered	
Is the population mean significantly different from a specified value?	
Model or structure	
Independent variable	A single numerical variable whose mean value is of interest.
Dependent variable	None
Equation model	$t = \dfrac{\bar{X} - \mu}{s_x/\sqrt{N}}$
Other structure	The P-value calculated from the t-score and the degrees of freedom, $N-1$, is the estimate of the probability that the sample mean would fall this far or further from the specified value, μ.
Corresponding nonparametric test	Wilcoxon signed rank test
Required assumptions	
Minimum sample size	20
Level of measurement	Interval
Distributional assumptions	Normal

TEST OF PROPORTIONS

As shown in Table 11-4, the one-sample z test of proportions is a statistical procedure that allows the estimation of the proportion of a population having some characteristic. This test can be used for categorical variables with two possible values.

The one-sample z test of proportions is useful in surveys and in process control. Suppose we want to introduce a new flavor to our line of soft drinks. We estimate that the additional flavor will be profitable if over 20% of our current customers like it. We take a sample of our customers and have them try the new flavor. The z test can tell us if the proportion of the population

Table 11-3   The paired *t* test.

Type of question answered	
Is the mean difference between scores taken from paired subjects different from zero?	
Model or structure	
Independent variable	Assignment to groups, one group to each member of pair.
Test statistic	Difference, D, between numerical scores of pair members.
Equation model	$$t = \frac{\sum D/N}{\sqrt{\sum(D - \bar{D})^2/N(N-1)}}$$
Other structure	The *P*-value calculated from the *t*-score and the degrees of freedom, $N-1$, is the probability that the observed difference would be this large or larger if there were no difference between the groups.
Corresponding nonparametric test	None
Required assumptions	
Minimum sample size	20 pairs
Level of measurement	Dichotomous/categorical for groups. Interval for scores.
Distributional assumptions	Scores must be normally distributed for both groups.
Other assumptions	Pairs must be well matched on extraneous variables or else linked on a prior basis (e.g., a married couple).

who will like the new flavor is significantly greater than $p = .20$. Or suppose we are manufacturing widgets and our contract with the customer commits us to less than a 1% rejection rate for quality. We can sample from the production line and use the z test to ensure that the population proportion of rejects is significantly below 1%.

Table 11-4   The one-sample z test of proportions.

Type of question answered	
Is the proportion of the population with a specific characteristic significantly different from a specified value?	
Model or structure	
Independent variable	A single dichotomous categorical variable.
Test statistic	The sample proportion, p, calculated as the number of individuals in the sample possessing the characteristic, divided by the sample size.
Equation model	$z \cong \dfrac{p_x - p}{\sqrt{p(1 - p)/N}}$
Other structure	The P-value calculated from the z-score is the estimate of the probability that the sample proportion would fall this far or further from the specified value, p.
Corresponding nonparametric test	The χ^2 test of proportions.
Required assumptions	
Minimum sample size	Both np and $n(1-p)$ must be greater than five.
Level of measurement	The independent variable must be dichotomous/categorical.

The same equations, used differently, allow us to calculate confidence intervals around a sample proportion. This means that we can take a survey and, depending on the sample size, give error bounds around the proportion of persons who respond in a certain way to a certain question. There is a big difference between a survey that says that $24 \pm 2\%$ of those surveyed use our product and one that says that $24 \pm 10\%$ do.

Correlation and Regression

This chapter covers the techniques involved in correlation and regression. Correlation and regression are ways of looking at data based on the scatter plot, which we saw in Figs. 7-12 and 7-13. The major difference between correlation and regression is that regression is a way of looking at causality. In regression, one set of variables (called the independent variables) are assumed to be possible causes. Another set (called the dependent variables) are assumed to be the possible effects. Using regression, the values of the independent variables for a specific individual can be used to predict the values of the dependent variable for that same individual. Correlation, on the other hand, is just a measure of the degree that higher or lower values on one variable have some correspondence to higher or lower values on another variable for a sample.

Relations Between Variables

The study of correlation and regression always begins with the simplest case, with just two variables measured for each individual in a sample drawn from the population. Later on, we will see how these relatively simple techniques can be expanded to deal with more variables (and the complexities that arise when we do).

INDIVIDUALS WITH CHARACTERISTICS

Key to understanding both correlation and regression is the underlying model of a population of individuals, each measured on a number of different variables. For any given individual, the values of those variables characterize that individual. We saw this in the example of Judy and her friends. Each person is characterized by height, weight, and I.Q. Of course, in a real study, there would be many more variables and, often, many more subjects. Note also that this model applies to both experimental and non-experimental studies. In an experimental study, we would have to distinguish carefully between variables measured before and after each intervention.

PLOTTING THE CORRELATION

Recall from Chapter 7 "Graphs and Charts" that we can draw the relationship between the values of two variables measured on the same subjects with a scatter plot. This is the geometric basis for the mathematics behind both correlation and regression. In Chapter 8 "Common Statistical Measures," we discussed a way of calculating the correlation coefficient that illustrated how it was a ratio of the variances relating the two variables. Here, we look at another way of calculating the same correlation coefficient that shows how it captures the geometry of the scatter plot. Here is another definition for the Pearson correlation coefficient:

$$r = \frac{\sum (x - \bar{X})(y - \bar{Y})}{(N-1)s_x s_y} \qquad (12\text{-}1)$$

Here, $\bar{X}$ and $\bar{Y}$ are the means of each of the two variables, and s_x and s_y are the standard deviations.

As we discussed in Chapter 8, standardizing values of a variable converts a normally distributed variable into a variable with a mean of zero and a standard deviation of one. As a matter of fact, standardization works with non-normally distributed variables. Standardization cannot make a non-normal distribution normal, but it will give it a mean of zero and a standard deviation of one. To standardize a variable, we take each value and subtract the mean of all the values and then divide by the standard deviation. Note that this new equation for the correlation coefficient shows its similarity to standardizing the product of the two values for each subject and adding them all together.

When the value of a variable is converted into a standard score, it becomes negative if it was below the mean and positive if it was above the mean. In terms of the sample, above the mean means high and below the mean means low. If two variables are directly related, when one value is high (or low), the other value will be high (or low) as well. In this case, most of the time, the standardized x-value and the standardized y-value will both be positive or both be negative, which means that the product will be positive. This will make the correlation coefficient higher. If the two variables are inversely related, when the value of one variable is high, the other will tend to be low, and vice versa. With the standardized values, most of the products will be negative and the correlation coefficient will be lower.

The relation between this formula and the geometry is illustrated in Fig. 12-1.

The points in the upper right-hand and lower left-hand portions will add to the correlation. The other points will lower the correlation. In Fig. 12-1, the correlation will be positive, with a value of .81, because most of the points fall in the places that raise the correlation above zero. If the two variables were unrelated, there would tend to be the same number of points in each of the four corners, and the correlation would be close to zero.

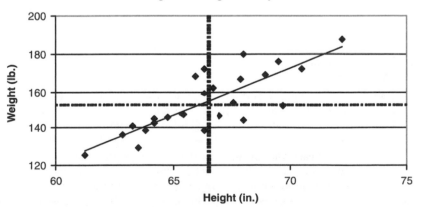

Fig. 12-1.   The geometry of correlation.

THE *t* TEST FOR THE CORRELATION COEFFICIENT

There is a one-sample significance test for the correlation coefficient. For reasons discussed below, we do not recommend it. We include it here because it is discussed in a number of business statistics texts.

Table 12-1   The *t* test for the correlation coefficient.

Type of question answered	
Is the correlation in the population significantly different from zero?	
Model or structure	
Independent variables	Two numerical variables measured on the same sample.
Test statistic	The Pearson product-moment correlation.
Equation model	$$t = \frac{r}{\sqrt{(1 - r^2)/(N - 2)}}$$
Other structure	The *P*-value calculated from the *t*-score and the degrees of freedom, $N-2$, is the probability that the observed correlation would be this far from zero or further if the true population correlation is zero.
Corresponding nonparametric test	Any of a number of alternative indices, including the Spearman rank-order correlation coefficient. (These are not presented in *Business Statistics Demystified*.)
Required assumptions	
Minimum sample size	20
Level of measurement	Interval (ordinal may not be used)
Distributional assumptions	Both variables must be normally distributed with equal variances. The conditional distributions of each variable dependent upon all values of the other variable must also be normally distributed with equal variances.
Other assumptions	The errors of each variable must be independent of one another. The values of each variable must be the product of random (not systematic) sampling. The true relationship between the variables must be linear.

CRITICAL CAUTION

There are a number of reasons to be very cautious in using the *t* test for the correlation coefficient. As we can see from Table 12-1, there are a number of assumptions, which, if violated, render the significance test invalid. While the test is moderately robust to violations of some of these assumptions, some of the assumptions, particularly the equality of variances for the conditional distributions, are often violated in real data. The linearity assumption can also be particularly troublesome, because it is an assumption about the relationship being measured. Many studies that use correlations involve either systematic sampling of at least one variable, or sampling procedures that create non-independent errors. The test is not very robust to these sorts of violations. Some texts even recommend restricted sampling over a range in which the relationship can be presumed linear, which violates the random sampling assumption in order to satisfy the linearity assumption.

There are two additional problems that relate to the meaning of the test itself. First, it is almost never the case in nature that two variables measured on the same subjects have a correlation precisely and exactly equal to zero, which is the null hypothesis for this test. This means that, given a large enough sample, every measured correlation will be significant! While this is a very general problem with any null hypothesis, it is especially troublesome for studies in which there is no intervention or control group, which is often the case in correlational studies. Furthermore, because interventions cost money, correlational studies tend to be larger than experimental studies, producing larger sample sizes. This gives rise to odd results, such as very small correlations that are statistically significant. What does it mean to say that two variables are significantly correlated with a coefficient of .01?

Second, the correlation coefficient is used when the relation between the variables cannot be assumed to be causal. If one of the variables is thought to be measuring the cause and the other the effect, regression tests, which have many advantages, can be used. The use of correlation instead of regression means that either we are ignorant of the true underlying causal relations, or we are unable to measure some additional variable or variables believed to be the cause of both of the two variables measured. In either case, the value of the correlation makes sense as a measure of the relationship found. However, the additional information that the correlation found is unlikely to be due to chance is difficult to interpret sensibly. All it really means is that we took a large enough sample, which is a fact entirely under our control and not reflective of anything about the nature of the data.

In an experimental study, the null hypothesis allows us to ask the question: Did our intervention have an effect? In a survey, the null hypothesis for the correlation only allows us to ask whether we collected enough data to get accurate measures of correlations of the size actually found, which is something we should have planned out in the first place.

EXERCISE

Note that, in the case of the heights and weights of Judy and her friends, we cannot assert that the correlation is significantly different from zero, despite the fact that we have a large enough sample and that the correlation is very large. As an exercise, say why the t test for the correlation coefficient cannot be used in this case.

CORRELATION AND CAUSALITY: POST HOC, PROPTER HOC

When two variables, A and B, are correlated, there are three standard possibilities. Either A causes B, or B causes A, or there is some third variable, C that causes both A and B. But the real world is much more complicated. Consider our simple example of height and weight. There is a sense in which being very tall necessitates having enough weight to fill out one's frame. The minimum weight for a short person is less than the minimum weight for a tall person. This fact could generate some correlation and probably accounts for part of the observed correlation. But is this truly a cause? Instead, we might say that there is a third characteristic of people, call it overall size, that is a possibly genetic cause of both weight and height. Perhaps, but certainly there are causes of height (like good diet and health in childhood) that are not causes of weight and vice versa. The values of every variable have multiple causes.

In addition, there is the problem of time. Our Western notion of causality includes the assumption that the cause must precede the effect in time. But our studies often do not measure a single subject across a long enough period of time to measure both causes and effects. Furthermore, many variables interact mutually over time, with increases in one leading to increases in the other, which lead to more increases in the first, etc. For example, if we track the number of employees of a company and its net worth, and both grow over time, it may well be that each is causing the other. The increase in net worth allows more hiring, and the larger workforce can be used to increase total sales, increasing net worth.

In all of these cases, both correlation and regression have very real limitations as techniques for assessing what is really going on.

Regression Analysis: The Measured and the Unmeasured

When we are in a position to assert that one or more variables measure causes and other variables measure their effects, we can use regression. The best case is when the independent variables measure the amount of intervention applied to each individual (such as fermentation time, weeks of training, or number of exposures to our advertisements) and the dependent variable measures change that would not be expected to occur without the intervention (such as sourness of the dough, number of sales, or amount of purchases). So long as certain additional assumptions are met, some form of regression analysis is the statistical technique of choice.

THE LINEAR REGRESSION MODEL

The most basic form of regression is simple linear regression. Simple linear regression is used in the case where there is one independent variable, X, presumed to measure a cause, one dependent variable, Y, presumed to measure an effect, and the relationship between the two is linear. In the scatter plot, the independent variable is graphed along the horizontal axis and the dependent variable is graphed along the vertical axis. We talk about the regression of X on Y.

When there are more variables, non-linear relationships, or other violations of basic assumptions, some other, more complex form of regression (discussed below) must be used. We will discuss simple linear regression in detail not because it is the most commonly used, but because it is the easiest to understand, and is the basis for all of the other forms.

What is a linear relationship?

Returning to Fig. 7-12, we see that a line has been drawn through the scatter plot. This line is called the *regression line*, and it is the heart and soul of the logic of regression analysis. While correlation attempts to summarize the relation shown in a scatter plot with a single number, regression attempts to summarize that same relation with a line. The rules for regression ensure that for every scatter plot there is one and only one "best" line that characterizes the cloud of points. That line defines an expected y-value for each x-value. The idea is that, if X causes Y, then knowing X should allow us to predict Y.

You may recall from algebra that any line can be expressed as an equation with two constants,

$$\hat{Y} = \beta_1 X + \beta_0 \qquad (12\text{-}2)$$

where β_1 is the slope of the line, describing how slanted it is, and β_0 is the y-intercept, indicating the point at which the line crosses the vertical axis when $X = 0$. Note that this means that whenever we know the value of X, we can calculate the value of $\hat{Y}$.

TIPS ON TERMS

We use the variable, $\hat{Y}$, instead of Y, because the points on our scatter plot are not in an exact line. $\hat{Y}$ is the variable that contains the values of our predictions of the y-values, not the exact y-values themselves.

Suppose we take one individual from our sample, let's use Francie, and look just at the x-value (height), and try to predict the y-value, weight. Figure 12-2 shows the scatter plot of heights and weights, with Francie's data point highlighted. Using the equation for the regression line to calculate a y-value is the way to use a regression analysis to estimate y-values. Geometrically, this is the same as drawing a vertical line from that x-value to the regression line, then drawing a horizontal line from that point on the regression line to the y-axis. The place where we hit the y-axis would be our estimate for Francie's weight. As we can see from Fig. 12-2, this procedure would give us an estimated weight of about 170 lbs for Francie, considerably above her actual

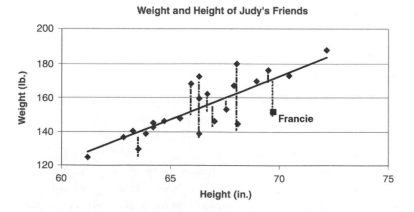

Fig. 12-2.   Regression residuals for weights and heights.

weight of 152. The vertical line from Francie's data point up to the regression line indicates the difference between her actual weight and the expected weight calculated by regression. It is called the *residual*.

The regression line is defined in terms of the residuals and the uniqueness of the regression line is determined by the values of the residuals. As it turns out, there is one and only one line that minimizes all of the residuals. That line is the regression line. If we use the regression line to predict *y*-values from *x*-values, we will do as well as we can for the items in our sample in the sense that our overall errors will be minimized.

KEY POINT

The regression line is the line passing through the data points that has the shortest possible total sum for the square of all the residuals. (This is called the *least-squares* criterion.) The regression line is unique for every set of points in two dimensions.

HANDY HINTS

Regression is Asymmetrical

Looking at Fig. 12-2, we note that, geometrically, the residuals for the regression of *X on Y* are all lined up parallel to the *y*-axis. Imagine that we are interested in the regression of *Y on X*. The scatter plot for this regression would be flipped around and the residuals would be parallel to the height axis instead of the weight axis. The lengths would be different and the regression line would not necessarily be the same.

In contrast, note that both of the equations for the correlation coefficient are symmetrical for *X* and *Y*. This means that, if we swap *X* and *Y*, the equation for the correlation coefficient does not change. This is because the correlation of *X* with *Y* is the same as the correlation of *Y* with *X*.

Causality is directional and so is regression. Of course, as we normally use regression, we put the possible cause, the intervention which occurred first in time, on the *x*-axis, calculating the regression of *Y on X*. Ordinarily, there is no reason to calculate the regression of *X on Y*, unless we wanted to claim a later event caused an earlier one.

There is, however, a precise mathematical relationship between correlation and regression. The Pearson product moment correlation is the slope of the regression line, adjusted for the difference in the standard deviations of the two variables. The correlation coefficient takes the differences in scale between the two variables into account in order to keep things symmetrical and to ensure that any Pearson product moment correlation for any two variables is scaled the same way. The regression line, on the other hand, is calculated for the values of the variables in their own original scales. The slope of the regression line is proportional to the correlation.

Significance in simple linear regression

Given that there is a regression line for every set of points, what does it mean for a regression to be statistically significant? Regression analysis is an attempt to find a causal relation. If there is a correlation, there may not be a causal relation, but if there is a causal relation, there must be a correlation. Therefore, we can use the absence of a correlation as our null hypothesis. This is the same significance test given in Table 12-1.

Another way of looking at this is that a significant regression means the ability to predict Y from X. The null hypothesis is that we cannot predict anything about Y from X. If X tells us nothing about Y, then being low or high on X has no effect on Y. A regression line where moving along the x-values does not change the y-values is horizontal. (Recall from algebra that the slope of a horizontal line is zero.) So, the appropriate null hypothesis is that the slope of the regression line is zero. Because the slope of the regression line is proportional to the correlation coefficient, if one is zero, the other is zero. So the two null hypotheses are equivalent.

As shown in Table 12-2, linear regression is a statistical procedure that allows the calculation of a significance level for the degree to which the values of one numerical variable (called the independent variable) predict the values of a second numerical variable (called the dependent variable).

REGRESSION ASSUMPTIONS

Note that the assumptions for regression given in Table 12-2 are basically the same as those for correlation in Table 12-1. The relevance of these assumptions is different because regression is intended to be used in the context of a controlled study where we have other reasons to assume a causal relation. In principle, it is possible to conduct a regression study with a large enough sample such that a very small correlation between the independent and dependent variable would be recorded as significant. However, there are separate statistics that can be used to evaluate the amount of error we can expect when we estimate the dependent variable. If our independent variable does not allow us to predict the dependent variable with sufficient accuracy

Table 12-2   Linear regression.

Type of question answered	
Can we predict values for one numerical variable from another numerical variable?	
Model or structure	
Independent variable	A single numerical variable assumed to measure a cause.
Dependent variable	A single numerical variable assumed to measure an effect.
Equation model	$\hat{Y} = \beta_1 X + \beta_0$
Other structure	The estimate of the slope, divided by the estimate of its standard error, is distributed as a t statistic. The equation is complex, but is equivalent to the equation in Table 12-1. The P-value calculated from the t-score and the degrees of freedom, $N-2$, is the probability that the observed slope would be this far from zero or further if the true population slope is zero.
Corresponding nonparametric test	None
Required assumptions	
Minimum sample size	20
Level of measurement	Interval
Distributional assumptions	Both variables should be normally distributed. The conditional distribution of the dependent variable at all values of the independent variable must also be normally distributed with equal variances.
Other assumptions	The errors of each variable must be independent of one another. The values of each variable must be the product of random (not systematic) sampling. The true relationship between the variables must be linear.

to be practically useful in the context of making our business decision, statistical significance is irrelevant.

In addition, there are also techniques (not covered here) that allow us to measure the degree to which various assumptions, particularly the linearity and independence of error assumptions, are violated. If there is any doubt about these assumptions, those tests should be performed.

Alternative types of regression

A special mention should be made of the linearity assumption. A causal relation may result in any one of an infinite number of systematic and important relations between two variables. Many of these relations are not linear. Recall from algebra that a linear equation is just the simplest of the polynomial equations. There are also quadratic equations, cubic equations, etc. Suppose low and high values of the independent variable lead to low values of the dependent variable, but middling values of the independent variable lead to high values of the dependent variable. For example, years of education is related to salary in this way. Up through college, increasing years of education tend to lead to increased income. But folks with Ph.D.s tend to make less money and bring down the average for everyone with more than 16 years of education. Or the situation may be reversed, with middling values of the independent variable leading to low values of the dependent variable. For example, one might find such a relationship between number of errors and team size. If a team is too small, the pressure to get all the work done would lead to errors. On a right-sized team, errors would decrease. When a team gets large, communication, training, and quality control are all more difficult, and we might find an increase in error again. These are very reasonable relationships and useful to know about. They are captured by quadratic equations and there are forms of regression analysis that allow us to assess them.

There are other forms of non-linearity that are not well handled by any polynomial function. In many cases, one or more of the variables can be transformed by some preliminary calculation so that the relation between these new variables is linear. Another form of non-linearity is when one relation holds for a particular range of *x*-values and another relation holds at other points along the *x*-axis. Complex forms of regression, using a technique called *splines*, are useful in these cases.

SOMETHING EXTRA

Get Ahead of the Curve—Use Splines

There is a marketing concept called the product life cycle. Total sales of a product start very slow, grow rapidly, drop at saturation, level off at maturity, and then drop to very low levels—or cease altogether—at obsolescence. An example might be total annual sales of new typewriters between the years 1880 and 2000. Traditionally, this is drawn with a smooth curve. The latest statistical techniques use splines—mixes of different lines and curves—to generate what some statisticians hope will be more accurate models. We might begin with an S-curve—slow start, exponential growth, leveling off at saturation. The mature phase might be a horizontal line, indicating flat sales. As typewriters entered obsolescence, probably about when Windows word processors with relatively high-quality printers came to consumers, we would see a steep S-curve for the decline to the point where few, or no, new typewriters are being sold every year.

Businesses plan very different survival and growth strategies based on their beliefs about the maturity of their market. Statisticians think splines will help. Be ready to use them!

When an independent variable is non-normally distributed, or even categorical, instead of numerical, regression analysis is relatively robust. Even dichotomous variables (called *dummy variables*) may be used. However, when a dependent variable is dichotomous, regression analysis is not robust with respect to this violation of distributional assumptions. Another complex, specialized type of regression, called *logistic regression*, can be used.

SURVIVAL STRATEGIES

The important thing to know is that there are many alternatives to simple linear regression that may serve our business needs. When in doubt, call on an expert to see if there are better ways to analyze the data.

While these other types of regression require other assumptions and are useful in other situations, the basic logic of simple linear regression applies to all of them. They are all attempts to characterize relationships between causes and effects in terms of mathematical functions. The shape of the function is always determined by the errors made in predicting the dependent variables. (There is one technical difference. For some types of regression more complicated than quadratic, cubic, or exponential, the least-squares method cannot be used and an alternative, called the *maximum likelihood* method, is used.)

Problems in prediction

Prediction is always a risky business. The large number of assumptions required for regression are an indication of this. In addition, there are specific problems in regression related to making predictions.

CRITICAL CAUTION

Predicting isn't Always About the Future

In statistics, prediction has many different uses. Relating to regression, it means determining the value of one variable for an individual from another variable or variables. It does not necessarily mean predicting the future. In fact, predicting the future, or forecasting, is a particularly difficult case of prediction.

In a regression context, making a prediction means taking an x-value that is not found in our sample, and calculating a y-value for that individual. The ability to make these sorts of predictions is very valuable in business, simply because measurement costs money. If we can measure just some of the variables and then calculate the rest, we can save money, time, and resources. If the number of contacts to current customers from salespeople predicts the number and value of sales by that customer, we can predict the optimal number of sales contacts to make per customer. This is an example where we would expect a nonlinear result. Up to a point, more sales contacts increase sales. Beyond that point, the customer may feel intruded upon, and sales may drop.

Of course, our prediction is just an estimate, based on our sample. There will be error. Furthermore, if the new individual is importantly different from those in our original sample, the prediction may go awry.

There is one way that new individuals may differ from those in our sample that can be easily measured. If the values of any of the independent variables for a new individual are outside the range of the independent variables found in our study sample, the prediction cannot be justified in a regression context. For example, none of Judy's friends are over six feet tall. If Judy makes a new friend who is $6'2''$, our prediction of this new friend's weight may not be valid.

TIPS ON TERMS

When we make a prediction for values of the dependent variable(s) based upon values of the independent variable(s) within the range of the sample values, the prediction is called an *interpolation*. When we make a prediction for values of the dependent variable(s) based upon values of the independent variable(s) outside the range of the sample values, the prediction is called an *extrapolation*.

The problems of extrapolation are particularly difficult in the case of forecasting. If our independent variable is time, then our predictions will always be extrapolations because our study is over and any new subjects will be measured in the future. The range of time used in our regression analysis is always in the past, because we took our sample in the past. A good example is predicting stock prices or our profits. Forecasting is always a battle against the problems of extrapolation. If these problems were easy to solve, we could all just play the stock market for a little while and then retire. We will discuss this in more detail in Chapter 16 "Forecasting."

Multiple Regression

Multiple regression (sometimes called multivariate regression) involves the use of more than one independent variable to predict the values of just one dependent variable. (In *Business Statistics Demystified*, we reserve the term "multivariate regression" for the much more complex situation where there are multiple dependent variables.) Here, we will discuss linear multiple regression only.

Earlier, we mentioned that we might predict income based on years of education. True, and we can get a much better prediction of income if we know years of education, age, race, family income of parents, marital status, and other factors. Having many such factors, and using them to increase the precision of our estimate, is very useful in business. Marketing companies sell statistical data using such factors to determine the likelihood of customers buying a company's product or service. Often the marketing companies provide general statistics organized by residential zip code to avoid giving away personal information about individual families. Although a corporate customer may use the data based on one variable—say, by mailing to selected residential zip codes—the value of the data lies in the fact that it aggregates a great number of variables about the population and their spending habits, and these multiple variables (per zip code, rather than per family) can be used to estimate the likelihood that people in a particular zip code are likely to buy the product. For example, we could go to a marketing company and say, "We know our product sells to young women between 14 and 17 years old in families with incomes over $50,000 per year. What zip codes have a large number of families in that income range with children that age?"

THE MULTIPLE REGRESSION MODEL

Statistically, multiple regression is a straightforward extension of simple regression. The chief advantage is that we are using more information about each subject in order to predict the value of the dependent variable. Multiple regression allows us to use many different measures to predict one. For example, we can use the customer's age, sex, income, type of residence, etc., to predict how much they will spend on an automobile. The use of multiple independent variables does create additional problems however. We will discuss these below.

As shown in Table 12-3, multiple regression is a statistical procedure that allows the calculation of a significance level for the degree to which the values

Table 12-3   Multiple regression.

Type of question answered	
Can we predict values for one numerical variable from multiple other numerical variables?	
Model or structure	
Independent variable	Multiple numerical variables assumed to measure causes.
Dependent variable	A single numerical variable assumed to measure an effect.
Equation model	$\hat{Y} = \beta_0 + \beta_1 X_1 + \beta_2 X_2 + \beta_3 X_3 + K + \beta_k X_k$
Other structure	The formula for testing the null hypothesis, which is expressed as a ratio of variances, is distributed as an F statistic. The equation is complex and is not covered here. The P-value calculated from the F-score and the degrees of freedom, $N-k-2$, is the probability that the observed slope would be this far from zero or further if the true population slope due to all variables is zero.
Corresponding nonparametric test	None
Required assumptions	
Level of measurement	Interval
Distributional assumptions	All variables should be normally distributed. The conditional distribution of the dependent variable at all values of all independent variables must also be normally distributed with equal variances.
Other assumptions	The errors of each variable must be independent of one another. The values of each variable must be the product of random (not systematic) sampling. The true relationship between the variables must be linear.

of more than one numerical variable (called the independent variables) predict the values of a separate numerical variable (called the dependent variable).

The null hypothesis for the F test in Table 12-3 is that there is no relation between the Y variables and any of the X variables. If any independent variable gives any information useful for predicting the dependent variable, the result will be significant. There is also a separate test, called the *partial F-test*, where the null hypothesis is that one independent variable contributes no additional information for the prediction beyond that provided by the other independent variables already in the model. The partial F-test is used in a number of complex procedures for deciding whether or not to include each of several candidate independent variables. There are different measures that can be used to make these decisions and they do not always give the same answers.

The issues of minimum sample size to establish significance are complex and a more advanced text (or an expert) should be consulted.

Any of the more complex forms of regression discussed in the preceding section on simple regression can also be part of a multiple regression model. In addition, there is a type of non-linear function specific to multiple regression models called an *interaction model*. This is where one independent variable has the effect of magnifying the effect of another. Interactions are very complex in a regression context, but a simple form found in a group context will be discussed in Chapter 13 "Group Differences."

MULTIPLE REGRESSION ASSUMPTIONS

All of the assumptions for simple regression apply to multiple regression as well. There is also the problem of collinearity. If some of the information contained in one independent variable useful in predicting the dependent variable is duplicated in another independent variable, then those two independent variables will be correlated. For example, salary, value of home, and years of education may all help predict the price of a person's car, but much of this information may reflect the amount of disposable income. In this sort of a case, we may get a good prediction of the dependent variable overall, but measures of the contribution of each independent variable to the prediction will be hard to determine. If we include salary first, the value of the home or the years of education may not make a significant contribution to the prediction, even though they may make a big contribution if included earlier.

Because there is no principled reason for including variables into the equation in any particular order and many variables are correlated to some degree, there is very often a problem with multiple regression in assessing the true contribution of any one variable. This can create very real problems in decision-making. For example, all studies that involve either the contribution of intelligence to some dependent measure, such as success, or which treat intelligence as a dependent measure and try to find out what makes folks smart, use a measure of the contribution to the regression called *percent variance accounted for*. All of these studies are subject to problems of collinearity. Despite this fact, proponents of these studies often propose serious policy decisions based on the notion that genetics determines intelligence, or intelligence determines success in life, and so forth.

The most conservative solution is simply not to take any measure of the relative contribution of any one independent variable too seriously. At a very minimum, genuine care must be taken to establish whether or not independent variables are correlated. Even with a study that includes only a few independent variables, other variables are not included in the study because they were too hard to measure or just not thought of, may be the real contributors.

Finally, we have the interventionist fallacy, also known as the Law of Unintended Consequences. Just because poverty leads to drug addiction does not mean that raising everyone's salary will lower the rate of drug use. Even if A causes B, changing the amount of A won't necessarily have the desired effect on B. The act of intervening may change the structure of the underlying causal relations.

CHAPTER

13

Group Differences: Analysis of Variance (ANOVA) and Designed Experiments

In Chapter 9 "Meaningful Statistics," we used the example of a group experiment to explain the concept of statistical significance. Here, we will cover the variations on group tests and discuss issues that arise from them. We will also show the relationship between group tests and regression.

Making Sense of Experiments With Groups

Recall from Chapter 9 that, historically, significance testing began with the notion of experiments where an experimental group received an intervention and a control group received none. Significance testing was designed to help make inferences as to whether or not the intervention had an effect measured in terms of a single, numerical dependent variable. Since that time, group testing has evolved to deal with many groups and multiple independent and dependent variables, similar to regression.

TIPS ON TERMS

When there are only two groups being compared, the statistical test used is called the *t* test, named after the statistical distribution used. When there are more than two groups, the statistical test is called ANOVA, short for the Analysis of Variance. The statistical distribution used is the *F* statistic.

While the underlying model for group tests is very different than for regression, it turns out that the underlying mathematics is identical, as we will see. The choice of which type of test to use is based on study design, not on any advantages of one technique over the other. In addition, regression has come to be more commonly used in non-experimental studies than have group tests. This is partly due just to tradition and partly due to the availability of many statistical measures in regression analysis for evaluating aspects of the data secondary to the overall significance. (These measures include ways of looking at the contribution of individual independent variables and even the influence of individual data points.)

WHY ARE GROUP DIFFERENCES IMPORTANT?

The main reason why group differences, and the group testing procedures used to analyze them, are important is that experiments with groups are the best way we know of to determine the effects of interventions. In business, we are often confronted with decisions as to whether or not to take some action. We most often want to make this decision based on the consequences of this action, its effects on profits, good will, return on investment, long-term survivability of our business, etc. If we can design an experiment (or quasi-experiment) to model this action as an intervention and then measure its effects, then the best way to analyze those effects is most often in terms of group differences.

THE RELATION BETWEEN REGRESSION AND GROUP TESTS

We mentioned in Chapter 12 "Correlation and Regression," that regression is robust if the independent variables are non-normal, even if they are ordinal/categorical. As it turns out, when all of the independent variables are categorical, regression procedures are mathematically identical to group tests. This is not a hard concept to see, at least in the simplest case.

Figure 13-1 shows a regression for Judy and her friends of height on sex. The diagram looks a bit silly because the independent variable, sex, is dichotomous. But notice that everything works out. The regression line is real. Its slope indicates that Judy's male friends tend to be somewhat taller than her female friends. If the regression is significant, that would mean that they are significantly taller.

The analysis shown in Fig. 13-1 is exactly equivalent to a t test of mean height between two groups, the men and the women. As it happens, when the independent variable is dichotomous, the regression line passes through the mean for each group. If the mean height for women is the same as the mean height for men, then the regression line will be horizontal (having a slope of zero). Thus, the null hypothesis for the t test, that the two means are the same, is identical to the null hypothesis of the regression, that the slope is

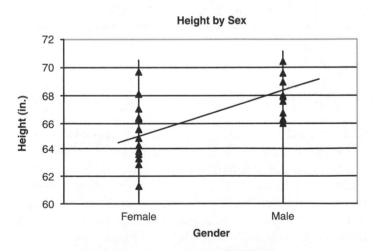

Fig. 13-1.   The geometry of group tests.

zero. In the case where there are more than two groups, the situation is more complex and we need to use an F test, but the regression and the group test are still mathematically equivalent tests of statistical significance.

Of course, in terms of our sampling procedures, we have not played strictly by the rules. We did not flip a coin to decide which of Judy's friends should have an intervention that would make them male. (Judy runs with a more circumspect crowd.) However, we have resolved the problem of the non-normal distribution of heights. The distribution of the dependent variable (height) is non-normal, but the two conditional distributions of height at the two values of the independent variable (female and male) are normal, which satisfies the normality assumption of the regression model.

EXERCISE

As an exercise, say why the regression line will pass through the mean of each group in Fig. 13-1. (Hint: Remember that the regression line is defined as the line that minimizes the residuals along the y-axis. Then remember the definition of the variance.) Will the regression line always pass through the mean of each group when the independent variable is ordinal, but has more than two values? If not, why not?

DESIGNS: GROUPS AND FACTORS

Let us consider the case where we regress a numerical variable on a categorical variable with more than two values. For example, we might regress the prices of a sample of used books on their condition, coded as: As New/Fine/Very Good/Good/Fair/Poor. In this case, we have a number of groups defined by their value on a single independent categorical variable. The variable is referred to as a *factor*. The values are referred to as *levels*.

Corresponding to multiple regression, we can regress a numerical variable on multiple categorical variables. In this case, we have multiple factors, each with multiple levels. Usually, this is pictured as a k-dimensional rectangular grid, with each dimension standing for one factor. Every group is defined in terms of a value for each categorical variable. For example, we could regress gas mileage on number of cylinders (4, 6, or 8), type of exhaust (with or without catalytic converter), and transmission (automatic, 4-speed, or 5-speed). Each car is assigned to one of the 18 groups based on its value for each of the three variables. Figure 13-2 shows this design.

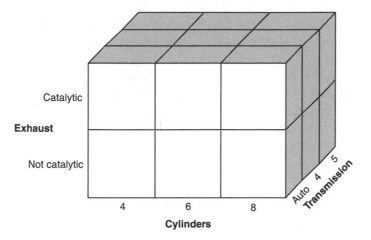

Fig. 13-2.   A 3-factor design.

HANDY HINTS

Note that, in the case of a categorical independent variable with more than two values, the significance tests for regression and group tests are equivalent even if the independent variable is just nominal and not ordinal. The reason is this: In general, the regression line does not pass through the mean of each group. And, should the means differ from group to group, the equation for the regression line does depend on the order of the groups. However, if the null hypothesis is true, the regression slope will be zero and the regression line will pass horizontally through the mean of all groups. In this case, the regression line will be exactly the same, even if the order of the groups is changed. Under the null hypothesis, the order of the groups does not matter, and should any of the groups have a different mean, the slope of the regression line will be non-zero, no matter what the order of the groups. In other words, the order of the groups affects the slope of the regression line, but does not affect whether or not the slope is zero, which is the only thing tested by the overall test of significance.

Group Tests

This section introduces the various statistical procedures for group tests.

COMPARING TWO GROUPS

The simplest group design is the two group design, which is analyzed using the *t* test. The two-tailed test asks if the group means differ in any way.

The one-tailed test asks if the experimental group mean is higher than the control group mean. (Alternatively, for the one-tailed test, we could ask if the experimental group mean is lower than the control group mean. What we cannot do is ask if it is either higher or lower. We must decide ahead of time which direction we care about. This is why the one-tailed test is said to use a *directional* hypothesis.)

As shown in Table 13-1, the *t* test is a statistical procedure that determines whether or not the mean value of a variable differs significantly between two groups.

COMPARING MORE THAN TWO GROUPS

The next simplest type of group test is when we have different varieties of intervention (called *treatments*). For example, we might assign our salespersons either one of two types of sales training, a motivation seminar, or no intervention. This would be a one-factor, four-level design. As shown in Table 13-2, the one-factor ANOVA test is a statistical procedure that determines whether or not the mean value of a variable differs significantly between multiple groups distinguished by a single categorical variable.

When more than two groups are studied, additional questions can be asked. The most common of these is whether two of the groups differ. Returning to our example, if we were to discover that our experiment with our sales staff had a significant effect, that would only mean that one of the treatments had produced a different mean. Any difference would be detected. For instance, it might be the case that the motivation seminar had lessened the performance of those salespersons. Overall significance for multiple groups by itself is often not helpful in making a business decision. If we find overall significance, we will want to see if one or more of the treatments has increased the mean performance of the sales persons beyond that of the control group. There is a separate *F* test for making these specific comparisons.

For reasons explained below, we will want to limit the number of specific comparisons made, especially if the decision as to which comparisons are desired is made after the data have been collected. After the overall significance test is done, we can examine the means to see what information about specific comparisons is most likely to help us make our business decision. For example, suppose the mean for one training program is much higher than all the other means. A comparison of this group to the control group is in order. But suppose that the group who took the motivation seminar (which happens to be much less expensive than either training program)

Table 13-1   The t test.

Type of question answered	
Does the mean of the dependent variable differ between two groups?	
Model or structure	
Independent variable	A dichotomous variable designating group assignment. Usually zero for the control group and one for the experimental group.
Dependent variable	A numerical variable measuring some quantity predicted to be affected by the differing treatments/interventions applied to each group.
Equation model	$$t = \frac{\bar{X}_1 - \bar{X}_2}{\sqrt{\left[\frac{(N_1 - 1)s_1^2 + (N_2 - 1)s_2^2}{N_1 + N_2 - 2}\right]\left[\frac{1}{N_1} + \frac{1}{N_2}\right]}}$$
Other structure	N_1 and N_2 are the sizes of the two groups. s_1 and s_2 are the sample standard deviations. The P-value calculated from the t-score and the degrees of freedom, N_1+N_2-2, is the probability that the observed difference would be this large or larger if there was no difference between the groups.
Corresponding nonparametric test	Wilcoxon rank sum test
Required assumptions	
Minimum sample size	20 per group
Level of measurement	Interval for dependent variable.
Distributional assumptions	Normal for dependent variable. Highly robust to violations.
Other assumptions	Random sampling for each group. Assignment of one individual to one group is independent of assignment of all other individuals to either group. (Random assignment to groups achieves this.) Group variances do not differ by more than a factor of three. Distribution of mean difference is normal.

Table 13-2 The one-factor ANOVA test.

Type of question answered	
Are the means of any of the groups unequal?	
Model or structure	
Independent variable	A single categorical variable designating group assignment.
Dependent variable	A numerical variable measuring some quantity predicted to be affected by the differing treatments/interventions applied to each group.
Equation model	Not applicable. The analysis of variance is described with a set of equations (not covered here) that relate differences in means between groups to two different variances: the variance of the means of the different groups (called the *between-groups* variance) and the variance of each score around the mean for its group (called the *within-groups* variance). These equations are designed so that if there is no true difference amongst the means, the two variances will be equal.
Other structure	The formula for testing the null hypothesis, which is expressed as a ratio of the two variances, is distributed as an F statistic. The P-value calculated from the F-score is the probability that the observed ratio would be this large or larger if the true group means were all equal. (Note that there are two separate degrees of freedom included in the ratio.)
Corresponding nonparametric test	Kruskal–Wallis
Required assumptions	
Minimum sample size	20 per group
Level of measurement	Interval for dependent variable.
Distributional assumptions	Normal within each group for dependent variable. (Moderately robust to violations.) The values of the independent variable must be predetermined.
Other assumptions	Random sampling for each group. Assignment of one individual to one group is independent of assignment of all other individuals to other groups. (Random assignment to groups achieves this.) Group variances do not differ by more than a factor of three.

also did somewhat better. We may want to see if this group did significantly better than the control group as well.

TWO FACTOR DESIGNS

When there is more than one type of intervention, we have a multiple factor design. This is equivalent to multiple regression in that there are multiple independent variables. The simplest case is when we have two interventions. We randomly assign individuals to four groups. The control group receives no intervention. Two of the groups receive one intervention each. The final group receives both interventions. This is called a two-by-two design.

For example, we might want to ask about the effects of both a sales training program and a motivational seminar. One group gets neither. One group gets sales training. One group gets the motivational seminar. One group gets both. Table 13-3 shows this design.

The advantage of a multi-factor design is that we can test to see if one independent variable has more or less of an effect depending on the level of some other factor. For example, perhaps the motivational seminar does not help untrained folks, but does improve the sales of those who have also received the sales training. This sort of effect is called an *interaction*. The other sort of effect we can test for is a difference between the means for just one factor. This sort of effect is called a *main effect*.

As shown in Table 13-4, the two-factor ANOVA test is a statistical procedure that determines whether or not the mean value of a variable differs significantly between multiple groups distinguished by two categorical variables.

Detecting interactions requires a great deal of statistical power. Often, even 20 subjects per group is not enough to detect important differences. In addition, if an interaction is found, the test for separate factors cannot

Table 13-3   Motivational seminar and training, a 2×2 design.

		Factor B Training	
		b_1 Did not receive	b_2 Received
Factor A Motivational seminar	a_1 Did not take	No intervention	Trained only
	a_2 Took	Motivated only	Trained and motivated

Table 13-4   The two-factor ANOVA test.

Type of question answered	
Are the means of any of the groups for any one factor unequal? Are the means for any factor affected by any combination of other factors?	
Model or structure	
Independent variable	Multiple categorical variables determining group assignments.
Dependent variable	A numerical variable measuring some quantity predicted to be affected by the differing treatments/interventions applied to each group.
Equation model	Not applicable. The analysis of variance is described with a set of equations (not covered here) that relate differences in means between groups to two different variances: the variance of the means of the different groups (called the *between-groups* variance) and the variance of each score around the mean for its group (called the *within-groups* variance). These equations are designed so that if there is no true difference amongst the means, the two variances will be equal.
Other structure	This design results in multiple F tests: one for each factor and one for the interaction. The formula for testing the null hypotheses, which are expressed as ratios of the two variances, are distributed as an F statistic. The P-value calculated from the F-score is the probability that the observed ratio would be this large or larger if the true group means were all equal. (Note that there are two separate degrees of freedom included in each ratio.)
Corresponding nonparametric test	None
Required assumptions	
Minimum sample size	20 per group
Level of measurement	Interval for dependent variable.
Distributional assumptions	Normal within each group for dependent variable. (Moderately robust to violations.) The values of the independent variable must be predetermined.
Other assumptions	Random sampling for each group. Assignment of one individual to one group is independent of assignment of all other individuals to other groups. (Random assignment to groups achieves this.) Group variances do not differ by more than a factor of three.

be relied upon. Check the interaction test first. If it is not significant, check the main effects for each factor.

MANY FACTORS, MANY GROUPS

The ANOVA test for two factors can be used for many factors. Separate F tests can be calculated for each factor and for every combination of factors. Things can get pretty confusing. The studies can also get very large, as individuals must be assigned to every group. Recall that, in our example of gas mileage, with just three factors, we had 18 groups. The number of individual subjects needed to achieve the needed statistical power can easily reach up to the hundreds. Big studies can be costly and that cost must be justified.

Specific comparisons between groups can be used with multiple factor designs, just as with one-factor designs. The problems associated with performing too many specific comparisons (discussed below) still apply.

Fun With ANOVA

Just as there are many additional types of regression not covered in *Business Statistics Demystified*, there are also many other types of ANOVA. Just as there are complexities that arise in larger regression studies, there are also issues with larger group studies.

BIGGER ISN'T NECESSARILY BETTER: THE PROBLEM OF MULTIPLE COMPARISONS

The problems associated with collinearity in regression are not a problem in ANOVA because separate F tests are used for each main effect and interaction. Instead, another problem arises: the problem of multiple comparisons. For main effects and interactions, the problem can be avoided, or at least skirted, by checking interactions first and then checking main effects only if the interactions are non-significant. This procedure is slightly more complex when more than two factors are involved. Higher-order interactions (interactions involving more factors) must be checked before lower-order interactions. For example, if we have three factors, A, B, and C, we must check the interaction of all three factors for significance first. If that is non-significant, we can check all three pairwise interactions next: A with B, B with C, and C with A. If all of those are non-significant, then we can check for main effects. The problem is much more serious when we deal with specific comparisons between groups.

We discussed the problem of multiple comparisons briefly in Chapter 3 "What Is Probability?" It is time for a little more detail and mention of some techniques used for solving it. The problem of multiple comparisons arises due to the fact that all statistical inferences involve probable events and, given enough attempts, even the most unlikely event is bound to occur eventually. In inferential statistics, we ensure conservatism by limiting Type I error to a small probability, the α-level, which is often set to .05. Suppose we had a large batch of random numbers instead of real data. By definition, there would be no real relations between these numbers. Any statistical test performed on random data that gives a significant result will be a Type I error.

However, if we performed 20 statistical tests of any variety on this random data, each with an α-level of .05, the odds are that one of the tests would give a statistically significant result, just because 20 times .05 is equal to one. We might perform all 20 tests and get no significant results, but eventually, if we kept on performing statistical tests on this random data, one or more would turn up significant, and false.

This is the same as rolling a pair of dice and trying to avoid a specific number coming up. The odds of rolling an eleven are one in eighteen, which is close to .05. Try rolling a pair of dice without rolling any elevens. See how far you get.

When we do a large study, whether it is a regression study or a group study, we are likely to have a lot of questions and want to perform a lot of tests. Even if there are no real relations in our data (equivalent to the case of having a big batch of random numbers), one out of every twenty tests is likely to come out significant. This undermines the principle of conservatism, which we must preserve if we are to have any justification for our conclusions.

The way statisticians deal with the problem of multiple comparisons is simple, but the details of the computations can get very complicated and we will not address them here. The solution that statisticians have adopted is to lower the α-level when multiple tests are performed on the same data or in order to answer related questions. There are many formulas for how much to lower the α-level for how many additional tests performed. One of the best and simplest is the Bonferroni, which is often available in standard statistical computer software. Unless a statistical consultant advises you otherwise, the Bonferroni technique is recommended.

One final note about multiple comparisons is that they are a specific case of the post hoc hypothesis, also discussed in Chapter 3. As such, the adjustment to the α-level required if we pick our specific comparisons before we collect our data is less than if we wait until afterwards. This is an important aspect to planning any large statistical study. If there are specific tests we anticipate will be of interest no matter what happens with our overall significance tests, we should plan them in advance and document them. This will allow us to use higher α-levels and get more statistical power.

ADVANCED TECHNIQUES

The various more advanced types of regression tend to focus on dealing with nonlinear relations between variables and variables that are scaled in ways that make it hard to detect linear relations. That is because regression is not especially robust to violations of these assumptions. For ANOVA, most of the advanced techniques address the definitions of the groups and the assignment of subjects to these groups, because ANOVA is not especially robust to violations of these assumptions.

Standard ANOVA techniques use what is called a fixed-effects model, because the levels of the factors are set by the experimenter. We choose which types of training to give to our sales people. In principle, however, there is a population of training techniques out there somewhere and the particular training techniques we are familiar with and have decided to evaluate in our study are a distinctly nonrandom sample from that population of training techniques. On occasion, we may select the levels of a factor randomly and must use a different type of ANOVA calculation to get the correct results. These ANOVA techniques use random-effects models. There are also techniques for studies where some factors are fixed and others are random. These are called mixed models.

Standard multi-factor ANOVA techniques require that all groups be the same size. More advanced techniques are available if there are different values of N for different groups. Critical to standard ANOVA is that subjects assigned to one group have no relation to any subject in any other group. Advanced techniques referred to as repeated-measures ANOVA allow for related subjects, or even the same subjects, to be assigned to different groups. Repeated measures, as the name implies, are very useful for measuring the effects of an intervention over time, such as at different points during an extended training program. There is also a version of ANOVA corresponding to multivariate regression, which uses multiple dependent measures. This is called MANOVA. MANOVA shares the same problems as multivariate regression in that the equations have multiple solutions.

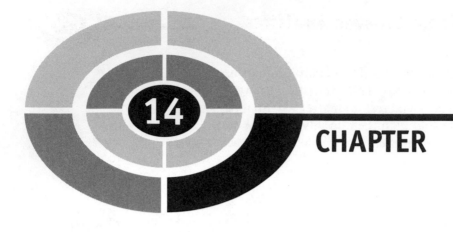

14

CHAPTER

Nonparametric Statistics

We discussed the theory behind nonparametric statistics in Chapter 9 "Meaningful Statistics." Here we will present some popular nonparametric tests. For more nonparametric tests, we recommend a book such as Mosteller and Rourke (1973).

Problems With Populations

As we discussed in Chapter 9, the most common reason to use a nonparametric test is when the appropriate parametric test cannot be used because the data do not satisfy all of its assumptions.

POORLY UNDERSTOOD POPULATIONS

It is rare that we have an extensive history of studies with the specific population we are studying. An exception is I.Q. The I.Q. test has been around for just over 100 years and the distribution of different I.Q. scores for different populations is well known. The distributions are normal and the means and standard distributions and standard errors are known and can be used safely. For almost every other sort of data, we need to look at our sample and test for or estimate the various characteristics of the population from which it is drawn. For example, in the manufacturing environment, we need to constantly monitor production processes with statistical process control, as we discuss in Chapter 17 "Quality Management." At any time, some factor could come in and change the mean, change the variance, or introduce bias to some important quality measure of our manufacturing process.

UNKNOWN POPULATIONS

Measurements are easy to make. We can measure physical characteristics of our products. We can ask questions of our customers. We can make calculations from our financial records, and so forth. It is not so easy to imagine the nature of the population from which our samples are drawn. When we lasso a few sheep from our flock, the flock is our population. But the flock is a sample as well. We can think of the flock as a sample of all living sheep, or of all sheep past, present, and future, or of sheep of those breeds we own, or even as a sample of all sheep we will own over time. Each of these populations is different and would be appropriate to asking different statistical questions.

For example, when looking at the relationship between color and breed, we should consider the population to be all sheep of that breed at the present time. It might be that, 100 years ago, a particular breed of sheep had many more black sheep than today. On the other hand our questions might have to do with the breeding patterns of our flock over time. We might want to know if our sheep have been breeding so as to increase or decrease the proportion of black sheep over time. In that case, our current flock is a sample from the population of all the sheep we have and will own over time.

Knowing the relationship between our questions and the theoretical population we are using is a crucial step in determining how to estimate the shape and the parameters of the population. Knowing the shape and parameters of the population are, in turn, critical in determining what sort of statistical test we can use.

A Solution: Sturdy Statistics

The questions we want to ask will narrow down our search for a statistical test. Parametric tests tend to answer only questions about parameters (such as the mean and the variance) of the population distribution. If we cannot phrase our question in terms of a population parameter, we should look to nonparametric tests to see if we can phrase our question in terms that can be answered by one of those. If our question can be answered by both parametric and nonparametric tests, the nature of the population will limit which tests we can use. If the assumptions required cannot be met for a parametric test, we can look to the corresponding nonparametric test, which is likely to have fewer assumptions that are more easily met.

REDUCING THE LEVEL OF MEASUREMENT

The most common nonparametric tests are used either when the level of measurement assumptions or the distributional assumptions cannot be met. These sorts of problems often come in tandem. For example, we may suppose that our customers' attitudes towards our products lie along some numerical continuum dictated by unknown psychological functions. How many levels of liking and disliking are there? Is there a zero point? What is the range of possible values?

In measuring attitudes, we ignore all of these rather metaphysical questions. Instead, we provide our subjects with a scale, usually using either five, or at most seven, levels, ranging from strongly dislike to strongly like. Most of the time, given a large enough sample size, attitudes measured in this way are close enough to being normally distributed that we can analyze the data with parametric tests. However, it may be the case that the limits of the precision of our measurement create very non-normal distributions. For example, if we are sampling current customers, it is unlikely that any will strongly dislike any of our products that they currently use. We might find ourselves with insufficient variance to analyze if, for instance, all of the customers either like or strongly like a particular product.

The good news is that such problems will be easy to detect, at least after the fact. A quick stem-and-leaf of our data will reveal narrow ranges, truncated distributions and other types of non-normality. (One solution is to pre-test our measurement instruments. We give the questionnaire to a small number of subjects and take a look at the data. If it looks bad, we consider rephrasing the questions.) On the other hand, if the data are already collected, there is not much we can do except to look for a statistical test that can handle what we have collected. The nonparametric tests designed as alternatives to parametric group tests, presented below, only assume an ordinal, rather than an interval, level of measurement. They often use the median, instead of the mean, as the measure of central tendency.

THE TRADEOFF: LOSS OF POWER

As we mentioned in Chapter 9 "Meaningful Statistics," the traditional tradeoff in choosing a nonparametric test is a loss of power. When the population is normal, the sample mean approximates the population mean closely with relatively small sample size. For other statistics, for other distributions, a close approximation takes a larger sample. As a result, tests that do not assume a normal distribution, or do not attempt to estimate the mean, tend to have less power.

Under these circumstances, the lowest cost solution is to pre-test our measurements and see if we can find a way to get numbers from our measurement techniques that are normally distributed. A little more effort in developing good measures can payoff in statistical power down the road. If we work at our measures, then we will only have to use lower-powered nonparametric tests when the population itself is non-normal, the mean is not a good measure of the central tendency, or the question we need answered cannot be answered in terms of the mean.

Popular Nonparametric Tests

There are many, many nonparametric tests. The most commonly used are tests of proportions, including tests of association, and rank tests, which replace common parametric group tests when their assumptions cannot be met.

DEALING WITH PROPORTIONS: χ^2 TESTS

As we discussed in Chapter 9, an important set of nonparametric tests are those that generate a statistical measure that is distributed as a χ^2. Like the t and F distributions, the χ^2 distribution is a theoretical curve that turns out to be the shape of the population distribution of a number of complex but useful statistics used in a number of different inferential statistical procedures. Its precise shape is known and, given the correct degrees of freedom, a P-value can be calculated.

Comparing proportions to a standard

Recall our example from Chapter 9 for using the χ^2 test to discover whether or not the breed of sheep in our flock affected the proportion of black sheep. The χ^2 test, more formally known as the Pearson χ^2 test, will provide an answer to this sort of question by determining whether the proportions in each individual data row are significantly different than the proportion across the summary total row at the bottom (called the column marginals).

Suppose we had a slightly different sort of question similar to our questions used in Chapter 11 "Estimation," in the test of proportions section, to illustrate the z test for proportions. Suppose that, instead of having to deliver ball bearings, 99% of which are to specification, we have a contract to deliver all of the production of our plant to a wholesaler, so long as the proportion of precision ball bearings, standard ball bearings, and bee–bees is 20/30/50. (All three items are manufactured by the same process, with a sorting machine that determines into which category the items fall.) Our wholesaler can only sell so many of each item. More precision ball bearings could be as problematic as more bee–bees. In order to ensure that we are shipping the right proportions, we sample a few minutes production every so often and obtain a count of each type of item.

We need to be able to determine if these three counts are in significantly different proportions from our contractual requirements. We can adapt the χ^2 test to this purpose by putting our data in the top row and then fake some data for the bottom row that corresponds exactly to the standard proportions required by our contract. In this case, our second row would just read: (*20 30 50 100*). We create the table with the totals as in Table 9-3. If the test is significant, then we know that we have not met the standard. The advantage to the χ^2 test over the z test is that we can use it for a multi-valued, rather than just for a dichotomous (two-valued) categorical variable. The disadvantage is that we cannot construct a true one-tailed test using the χ^2. Using the χ^2, we cannot ask if the actual proportions fail to meet the standard, only if they are different from the standard (either failing it or exceeding it).

As shown in Table 14-1, the χ^2 test for proportions is a nonparametric statistical procedure that determines whether or not the proportion of items classified in terms of a categorical variable (with different values in different columns) differs from some fixed standard proportion.

Tests of association

In Chapter 9 "Meaningful Statistics" and also in Chapter 12 "Correlation and Regression," we discussed the difference between relations between variables in general and those that are due to underlying cause–effect relationships. When a relationship between two variables is not assumed to be one of cause and effect, the relationship is usually measured using some sort of correlation coefficient. Some statisticians think of a correlation as something specifically measurable by the Pearson product moment correlation and use the term *association* for a general, non-causal relation. When

Table 14-1 The χ^2 test for proportions.

Type of question answered	
Do the proportions of a mixture of items differ from a standard set of proportions?	
Model or structure	
Independent variable	A categorical variable containing counts of each item falling into one of c categories.
Required calculation	The expected value for each cell in the table, E_{jk}, calculated as the row marginal times the column marginal, divided by the grand total.
Equation model	$$\chi^2 = \sum_j \sum_k \frac{(O_{jk} - E_{jk})^2}{E_{jk}}$$
Other structure	For the top row, $j=1$, the observed values in the cells, O_{jk}, are the data. For the bottom row, $j=2$, the observed values in the cells, O_{jk}, are the integer values corresponding to the standard proportions. The P-value calculated from the χ^2-value, with degrees of freedom, $(c-1)$, is the estimate of the probability that the sample proportion would fall this far or further from the specified standard.
Corresponding parametric test	One-sample z test of proportions
Required assumptions	
Minimum sample size	5 per cell
Level of measurement	Nominal/categorical
Distributional assumptions	None

dealing with two categorical variables, the Pearson product moment correlation cannot be used. Instead, there are a very large number of measures used depending upon the type of question being asked and the assumptions made about the two categorical variables and the relationship between them. These measures are universally called *tests of association*.

Understanding tests of association involves understanding how the notions of dependence and independence, discussed in Chapter 3 "What Is Probability," apply to categorical variables. Recall that our notion of statistical dependence relied upon the idea that knowing the value of one variable as providing information as to the probability of the value of another variable. In terms of our example of breeds and colors of sheep in Table 9-3, a dependency would mean that knowing which breed of sheep a particular sheep is, will tell us something about what color it is likely to be (or vice versa). If each breed of sheep has exactly the same proportion of black and white sheep, then knowing the breed tells us nothing more about the likelihood that the sheep is black than does the overall proportion of black sheep in the flock. So a test of independence is designed to ask if the proportions in one row or column differ significantly from another. A difference in proportions means information and dependence. Conveniently, this is exactly the same question as for our χ^2 test of proportions, so the calculations are identical. When we test two variables for independence, we are also testing to see if the proportions in the rows differ from column to column, and vice versa.

TIPS ON TERMS

Contingency table. A table showing the relationship between two categorical variables. Each cell contains the count of items with the corresponding values for each variable. The totals for each row and each column, plus the grand total are calculated. A theoretical contingency table contains proportions in each cell, with a grand total of one.

CRITICAL CAUTION

It is extremely important to note that the calculations for the Pearson χ^2 test, which is the only test of association we will cover here, are symmetrical for rows and columns. In other words, in Chapter 9 "Meaningful Statistics," had we elected to discover whether the color of sheep affected the breed (an admittedly odd way of looking at things), we would get the exact same numbers and the exact same results. The χ^2 test looks for any statistical dependencies between the rows and the columns, in either direction.

As shown in Table 14-2, the χ^2 test of independence is a nonparametric statistical procedure that shows whether or not two categorical variables are statistically independent. This test is equivalent to the Pearson χ^2 test for comparing proportions, which is a nonparametric statistical procedure that determines whether or not the proportion of items classified in terms of one

Table 14-2   The χ^2 test of independence.

Type of question answered	
Is one variable related to another?	
Model or structure	
Independent variable	Two categorical variables, the first containing counts of each item falling into one of r categories and the second having c categories.
Required calculation	The expected value for each cell in the table, E_{jk}, calculated as the row marginal times the column marginal, divided by the grand total.
Equation model	$$\chi^2 = \sum_j \sum_k \frac{(O_{jk} - E_{jk})^2}{E_{jk}}$$
Other structure	The P-value calculated from the χ^2-value, with degrees of freedom, $(r-1)(c-1)$, is the estimate of the probability that the sample proportion would fall this far or further from the specified standard.
Corresponding parametric test	None
Required assumptions	
Minimum sample size	5 per cell
Level of measurement	Nominal/categorical
Distributional assumptions	None

categorical variable (with different values in different columns) is affected by the value of another categorical variable (with different values in different rows).

Two notes on the calculations: The value of the Pearson χ^2 test statistic is higher when the observed cell values, O_{jk}, differ more from the expected cell values, E_{jk}. The observed cell values are just the data. The expected cell values are the counts that would be in the cells if the two variables were independent and there were no error. The equation given for calculating the expected cell values, E_{jk} (listed as *Required calculation* in Table 14-2), is much simpler than it appears. We assume that all of the totals are the same and use the totals to calculate the cell values in reverse. If the variables were truly independent, then all of the proportions in all of the rows would be equal, as would all of the proportions in all of the columns. The proportion for a row (or column) is just the row (or column) total divided by the grand total. The count for any one cell is just the total for that column (or row) times the proportion for the row (or column).

The degrees of freedom, $(r-1)(c-1)$, is based on the size of the table, $r \times c$. We subtract one from the number of rows and one from the number of columns because we have used up these degrees of freedom when we calculated the totals. One easy way to think about this is to take a look at an almost empty 2×2 table with the totals calculated:

Table 14-3　Degrees of freedom for the χ^2 test: Sheep by color and type of wool.

Sheep by Color and Type of Wool			
	White	Black	Total
Heavy wool	42		48
Fine wool			90
Total	118	20	138

EXERCISE

The number of degrees of freedom for the χ^2 test of a contingency table is the same as the number of cells that can be varied freely without altering the totals.

The number of degrees of freedom for a 2×2 table is $(2-1)(2-1)=1$. One cell in Table 14-3 is filled in. This means that none of the other three cells can vary. As an exercise, use the totals to calculate the counts that must go into the other three cells. Do not refer to Table 9-2.

Estimating the population variance

Sometimes, we need to know about the variance of a population, instead of the mean. While this is technically the estimation of a parameter of a normal distribution and, as such, is a parametric procedure, but the test statistic is distributed as a χ^2, so we cover it here. As shown in Table 14-4, the χ^2 test for population variance is a parametric statistical procedure that evaluates an estimate of the population variance with respect to some specific value.

Table 14-4   The χ^2 test for population variance.

Type of question answered	
Does the sample variance differ significantly from a specified value?	
Model or structure	
Independent variable	A single numerical variable whose variance is of interest
Dependent variable	None
Equation model	$$\chi^2 = \frac{(N-1)s^2}{\sigma^2}$$
Other structure	The P-value calculated from the χ^2-value and the degrees of freedom, $N-1$, is the estimate of the probability that the sample variance would fall this far or further from the specified value, σ^2.
Corresponding nonparametric test	None
Required assumptions	
Minimum sample size	20
Level of measurement	Interval
Distributional assumptions	Normal

As we will see in Chapter 17 "Quality Management," the proportion of precision ball bearings, standard ball bearings, and bee–bees in our ball bearing example is dependent upon the variance of the diameter of the balls produced. To ensure that the desired proportions are being manufactured, we could monitor the production line using statistical process control and test to see if the variance differed significantly from the desired variance.

ALTERNATIVES TO *t* TESTS: WILCOXON RANK TESTS

Among the many available nonparametric tests are two tests that can be used in place of *t* tests when the population distribution is so non-normal that the *t* test is not robust. Both tests were developed by Wilcoxon, both use the median instead of the mean, and the calculations for both involve ranking the data.

Ranking data is a common technique in nonparametric testing. When the numerical values for a variable are derived from an unknown or radically non-normally distributed population, the precise numerical values do not provide especially useful information about the central tendency. By renumbering all the data points with their ranks, we actually lessen the amount of information in the data, but we retain all the information needed to estimate the median. So long as the median is a reasonable measure of the central tendency (which is true for roughly symmetrical distributions), ranking provides a convenient means of generating a test statistic from which a *P*-value can be calculated.

Single-samples: the signed rank test

As shown in Table 14-5, the Wilcoxon signed rank test is a nonparametric statistical procedure that evaluates an estimate of the population median with respect to some specific value. The details as to how to perform the calculations for this procedure are a bit cumbersome, and are covered in most textbooks on business statistics.

Two groups: the rank sum test

As shown in Table 14-6, the Wilcoxon rank sum test is a nonparametric statistical procedure that determines whether the difference between the medians of two groups is significant. It is a good replacement for the *t* test when the population distribution is non-normal. It works for ordinal data with less than five levels.

MULTI-GROUP TESTING: THE KRUSKAL–WALLIS TEST

The Kruskal–Wallis test is to one-factor ANOVA as the Wilcoxon rank sum test is to the two-group *t* test. As shown in Table 14-7, the Kruskal–Wallis test is a nonparametric statistical procedure that determines whether the difference between the medians of several groups is significant.

Table 14-5   The Wilcoxon signed rank test.

Type of question answered	
Is the population median significantly different from a specified value?	
Model or structure	
Independent variable	A single numerical variable whose mean value is of interest
Dependent variable	None
Equation model	$z = \dfrac{W - \left(N'\left(N' + 1\right)/4\right)}{\sqrt{(N'(N' + 1)(2N' + 1)/24)}}$
Other structure	First, the data values are converted to difference scores by subtracting the median and taking the absolute value. Any scores of zero are omitted and the number of non-zero scores, N', is used in place of N. Ranks are assigned and the positive and negative signs are put back in. The Wilcoxon value, W, is the sum of the positive ranks. For $N < 20$, the P-value is calculated directly from the Wilcoxon score and the value of N. For larger sample sizes, a z-score is calculated as given in the equation above. In both cases, the P-value is the estimate of the probability that the sample median would fall this far or further from the specified value, M.
Corresponding parametric test	One-sample t test
Required assumptions	
Minimum sample size	5
Level of measurement	Interval
Distributional assumptions	Any roughly symmetrical distribution
Other assumptions	The true population values are continuous.

Table 14-6   The Wilcoxon rank sum test.

Type of question answered	
Does the median of the dependent variable differ between two groups?	
Model or structure	
Independent variable	A dichotomous variable designating group assignment. Usually zero for the control group and one for the experimental group.
Dependent variable	A numerical variable measuring some quantity predicted to be affected by the differing treatments/interventions applied to each group.
Equation model	$$z = \frac{T_1 - \mu_{T_1}}{\sigma_{T_1}}$$
Other structure	Ranks are assigned to all values, irrespective of group. The test statistic, T_1, is the sum of the ranks of the data points in the first group. (The first group is the smaller group if the Ns are unequal.) For $N < 20$, the P-value is calculated directly from the values of T_1 and N. For larger sample sizes, a z-score is calculated as given in the equation above. In both cases, the P-value is the probability that the observed difference would be this large or larger if there were no difference between the groups.
Corresponding parametric test	The t test (for independent groups).
Required assumptions	
Minimum sample size	8 (4 per group)
Level of measurement	Ordinal
Distributional assumptions	None

Table 14-7   The Kruskal–Wallis test.

Type of question answered	
Are the medians of any of the groups unequal?	
Model or structure	
Independent variable	A single categorical variable designating group assignment for k groups.
Dependent variable	A numerical variable measuring some quantity predicted to be affected by the differing treatments/interventions applied to each group.
Equation model	$$H = \left[\frac{12}{N(N+1)} \sum \frac{T_j^2}{N_j} \right] - 3(N+1)$$
Other structure	Ranks are assigned to all values, irrespective of group. N_j is the size of the jth group. T_j is the sum of the ranks assigned to the jth group. The P-value calculated from the H-score and the degrees of freedom, $k-1$, is the probability that the observed difference would be this large or larger if there were no difference between the groups.
Corresponding parametric test	One-factor ANOVA test
Required assumptions	
Minimum sample size	5 per group
Level of measurement	Ordinal
Distributional assumptions	If the group populations are assumed to have the same shape and variability, the null hypothesis is that the medians are equal. If not, the null hypothesis is that there is no difference between the distribution of the groups.
Other assumptions	The true population values are continuous.

Exam for Part Three

1. Which of the following is *not* a property of the mean?
   (a) It is consistent
   (b) It is unbiased
   (c) It is sufficient
   (d) All of the above

2. A 95% confidence interval is the interval in which the _____ value is 95% likely to fall.
   (a) Estimated
   (b) True population
   (c) Confidence
   (d) Error

3. When we have no independent source of information as to the variance of the population, we must use our best estimate of the population variance. That is the _____.
   (a) Confidence interval
   (b) Error
   (c) Sample variance
   (d) None of the above

4. In terms of the type of question answered, the _____ test is a group test.
   (a) Paired t
   (b) z
   (c) Mean
   (d) Population variance

5. The one-sample Z test of proportions is useful for calculating _____ around a sample proportion.
   (a) Difference test
   (b) Confidence interval
   (c) Both (a) and (b)
   (d) Neither (a) nor (b)

6. The geometric basis for the mathematics behind both correlation and regression is the _____.
   (a) t test
   (b) Z test
   (c) Bar graph
   (d) Scatter plot

7. When the value of the variable is converted into a standard score, it becomes _____ if it is below the mean and _____ if it is above the mean.
   (a) Positive; positive
   (b) Positive; negative
   (c) Negative; positive
   (d) Negative; negative

8. The correlation coefficient's assumption of _____ is often violated in real data.
   (a) Equality of variances for the conditional distributions
   (b) Linearity
   (c) Independent errors
   (d) All of the above

9. Which of the following is a standard possibility when 2 variables, A and B, are correlated?
   (a) A causes B
   (b) B causes A
   (c) A third variable causes both A and B
   (d) All of the above are standard possibilities

10. When drawing a regression line, you want to _____ the residuals.
    (a) Maximize
    (b) Minimize
    (c) Estimate
    (d) Predict

11. The problems of _____ are particularly difficult in the case of forecasting.
    (a) Extrapolation
    (b) Interpolation
    (c) Both (a) and (b)
    (d) Neither (a) nor (b)

12. Multiple regression allows us to use many different _____ variables to predict one _____ variable.
    (a) Dependent; dependent
    (b) Dependent; independent
    (c) Independent; dependent
    (d) Independent; independent

13. The problem of _____ in multiple regression is not found in simple regression.
    (a) Equality of variances for the conditional distributions
    (b) Collinearity
    (c) Independent errors
    (d) Error

14. Experiments with groups are the best way to determine the _____.
    (a) Effects of interventions
    (b) Correlation coefficient
    (c) Regression line
    (d) All of the above

15. The null hypothesis for a _____ test is that the difference between groups is exactly zero.
    (a) One-tailed
    (b) Two-tailed
    (c) Both (a) and (b)
    (d) Neither (a) nor (b)

16. Regression procedures are mathematically identical to group tests when all of the _____ variables are categorical.
    (a) Null
    (b) Error
    (c) Dependent
    (d) Independent

17. In designs, variables are referred to as _____, and values are referred to as _____.
     (a) Levels; levels
     (b) Levels; factors
     (c) Factors; levels
     (d) Factors; factors

18. The _____ test is a statistical procedure that determines whether or not the mean value of a variable differs significantly between multiple groups distinguished by two categorical variables.
     (a) Two-factor ANOVA
     (b) One-factor ANOVA
     (c) Correlation
     (d) Regression

19. With ANOVAs, the problem of _____ may arise.
     (a) Main effects
     (b) Interactions
     (c) Multiple comparisons
     (d) Collinearity

20. If we cannot phrase our question in terms of a population parameter, we should look to _____ tests.
     (a) ANOVA
     (b) Regression
     (c) Parametric
     (d) Nonparametric

21. Nonparametric tests are _____ powerful than parametric tests.
     (a) More
     (b) Less
     (c) Equally

22. The _____ is a statistical procedure that can determine whether the proportion of items classified as a categorical variable differs from some fixed standard proportion.
     (a) χ^2 test of proportions
     (b) χ^2 test of population variance
     (c) χ^2 test of independence
     (d) All of the above

23. The _____ is a statistical procedure that can evaluate an estimate of the population variance with respect to some specific value.
   (a) χ^2 test of proportions
   (b) χ^2 test of population variance
   (c) χ^2 test of independence
   (d) All of the above

24. The _____ is a statistical procedure that evaluates an estimate of the population median with respect to some specific value.
   (a) χ^2 test of independence
   (b) Wilcoxon rank sum test
   (c) Wilcoxon signed rank test
   (d) Kruskal–Wallis test

25. The _____ is a nonparametric statistical procedure that is similar to the one-factor ANOVA.
   (a) χ^2 test of independence
   (b) Wilcoxon rank sum test
   (c) Wilcoxon signed rank test
   (d) Kruskal–Wallis test

PART FOUR

Making Business Decisions

Part Four of *Business Statistics Demystified* is about *Making Business Decisions*. Here, we take all the statistics we've learned and apply it to different areas of business decision making and management. Of course, statistics is actually applied to dozens of areas. In a short introductory text, we chose to focus on three illustrative examples that show the most important issues in applying statistics to business, and also a sample of the very different ways in which statistics is used in business.

In Chapter 15 "Creating Surveys," we look at a common challenge for the young manager applying statistics in marketing, and also in HR, designing and conducting an opinion poll. Chapter 16 "Forecasting" shows why predicting the future is such a challenge, and what statisticians have to offer in that regard. In Chapter 17 "Quality Management," we show how statistical techniques have moved from the manufacturing environment into the realm of business processes, supporting decisions that create real improvements in the bottom line.

CHAPTER 15

Creating Surveys

Here, we focus on how to design and carry out surveys in an effective and statistically valid manner so that our business surveys will provide information we can use to support business.

The need for expertise and experience in surveys is far more critical even than in other areas of business statistics. A poorly designed or executed survey can do irreparable damage to our business. The detailed work of setting up even the simplest survey is very easy to do, but very hard to know how to do correctly. The guidelines below are for opinion polls. All of the rules for other kinds of surveys apply to opinion polls, and opinion polls have important rules of their own. The steps in performing a survey are: Planning and design, conducting the survey, and interpreting and reporting the results.

Planning and Design

To plan our survey, we must define our goals and objectives, identify our target population, choose our data-gathering methods, pick a sampling procedure, and write the survey questions.

STATE GOALS AND OBJECTIVES

As always, we start with our business questions. What information will be useful in answering them? By the time our plan is in draft form, our questions must be phrased clearly and precisely. And the questions must be clear to the people we interview, not just to us and our team.

We need to state our objectives clearly and precisely and most of all simply. What is the goal of the survey? State it in one sentence. What are the sub-goals? Make sure that everyone signs off on all objectives and the wording describing them.

IDENTIFY TARGET POPULATION

As in an experimental study, we are going to take a sample and make inferences about a population. All of the concerns discussed earlier in *Business Statistics Demystified* apply here as well. In addition, in a survey, the choice of a target population has enormous impacts on our survey method. Do we want to find out about customers or potential customers or competitors? The way we contact our respondents and obtain information from them may depend critically upon who they are and what sort of relationship we have with them (if any).

CHOOSE METHOD(S)

There are four basic types of survey method:

- *Mail-out surveys.*
  - *Advantages:* This is the least expensive method (except for online Internet surveys). It is convenient for the respondent and allows him or her ample time to reply. The questionnaire, any supporting documents, and the envelopes (in the case of snail-mail) can be printed so as to make clear that we are who we say we are and that we are established, credible, and responsible. The absence of an interviewer gives the respondent a greater sense of privacy.
  - *Disadvantages:* This is the slowest method. It also has the lowest average response rate, which means lots of follow-up mailings to get a large enough sample. We can never be sure that it was our target who actually filled out the questionnaire. Secretaries and assistants answer a lot of mail-out surveys. If the respondent doesn't understand the question, there is generally no one to ask. Most importantly, there is a strong self-selection bias to the sample.

- *Telephone surveys*. Potential respondents are contacted by interviewers by telephone.
  - *Advantages:* This is the fastest method. Large numbers of respondents, spread over a very wide area, can be contacted in a very short time, and, by using professional telephone survey banks, at very low cost. (Automated telephone surveys can be even less expensive, but these seem to have lower response rates and the advantages of having a live interviewer are lost.) The interviewer can control the order in which questions are asked and make sure the respondent understands them.
  - *Disadvantages:* Before almost everyone had a telephone, there was a built in sampling bias to the telephone survey. Now that some people are abandoning their landlines for cell phones, this bias is re-appearing. Even if it becomes legal to survey by calling cell phones, cell phones sample people and not places, so the two methods may not be easy to mix. No visual materials can be presented. And, of course, the respondent can always end the interview prematurely by hanging up.
- *In-person interviews*. The interviewers meet with each respondent in person.
  - *Advantages:* This method can provide the most detailed data. The interviewer may have pictures, charts, and especially product samples. (This is the only method when we want to do a taste test.) It may be the only way to reach hard-to-reach populations, such as the homeless, incarcerated prisoners, or CEOs. This method also has the highest response rate, which can mean less sampling bias.
  - *Disadvantages:* This is the most expensive method. The personal presence of the interviewer increases the likelihood of bias introduced by the individual psychological makeup or personal biases of the interviewer. The interview is not as safe for the interviewer, nor as private for the respondent, as other methods.
- *Online (Internet) surveys*. Potential respondents are invited to visit a website where the questionnaire is presented by the computer. Online surveys are too new for their advantages and disadvantages to be well understood, but a few things are clear:
  - *Advantages:* Potentially, this method can combine an even lower-cost than mail-out surveys with some of the flexibility of telephone surveys. Graphics, and even video and audio are available. If we sell our products online, we can reach our customers at the point and time of purchase.

- *Disadvantages:* Only computer users can be reached. Snazzy graphics and audio–visual materials will only run on some computers. Skillful programming will be needed to provide the flexibility of ordering the questions depending upon the response. Computers are not interviewers. There is only a limited ability to clarify a respondent's questions. Computer graphics are not print graphics. They have neither the quality, nor convey the authority, of print graphics. Most importantly, sampling bias may be created by the method by which respondents are invited to participate. The so-called *unscientific* online surveys popular on TV news and entertainment shows are barely surveys at all.

In order to get the sample we want, we may use one method initially and follow-up with others.

PICK THE SAMPLING PROCEDURE

The order of the above elements of planning and designing our survey is not locked in stone. However, we must state goals and objectives so that we can identify our target population, and then make good decisions about how to reach our sample. Picking the sampling procedure and choosing our overall survey method are closely interlinked. For example, initial random sampling by phone or by mail is possible, but different sorts of bias are introduced by the choice of method. Also, sample size may be constrained by the costs of one method over another.

Survey sampling is very complex, far more so than other types of sampling we have discussed in *Business Statistics Demystified.* (Deming's, 1950, non-comprehensive classic on survey sampling, *Some Theory of Sampling*, is 602 pages long, and it is over 50 years old.) Expert advice should be sought in all cases. In order to be an intelligent consumer of that expert advice, we recommend a class in survey research, or a good book, such as Rea and Parker's (1997) *Designing and Conducting Survey Research.*

The important thing to understand about survey sampling is that, we may want to know how many demographic variables, such as age, sex, economic status, etc., relate to our respondents' opinions. Ensuring that our sample is correctly representative of our population on all of these variables is no easy matter.

WRITE THE SURVEY QUESTIONS

The most difficult and important part of any survey is writing the questions. There are even entire books about writing effective survey questions, such as Converse and Presser (1986). But no book can teach everything about writing good survey questions. Experience is the key. If no one in our organization has the experience, we need to hire someone who does. We can draft our own questions, of course, but an expert should edit them before we use them.

Here are some different types of questions:

- *Dichotomous.* A yes–no or true–false question.
- *Multiple-choice.* Questions that allow a limited number of predefined answers, usually measured as a nominal variable.
- *Ratings scale.* A question answered with values on an ordinal scale. For example: Strongly disagree. Disagree. No opinion. Agree. Strongly agree.
- *Completion.* A fill in the blank question. Used most often for nominal variables with large numbers of possible options. Example: What is your occupation?
- *Open-ended.* A question that allows the respondent to answer freely. This type of question provides the most information, but the answers are hardest to analyze.

Except for open-ended questions, all questions should be carefully phrased so that as many of our respondents as possible can choose one of the available alternatives. For example, we can offer many options for selection, and then leave an "other" field to get the benefits of both multiple-choice and completion questions.

Here are some guidelines for writing good questions:

- *Reduce error*
  - Write short questions
  - Avoid double-negatives
  - Use simple language
  - Use simple concepts. Avoid abstract, complex, technical, specialized, or otherwise unfamiliar language.
  - Be specific and avoid ambiguous expressions
  - Be personal. Whenever possible, ask about the respondent's own experience and knowledge.
  - Be current. Most people have poor memory. Ask about recent events. Tie less recent times to specific, memorable occasions, rather than to dates.

- *Reduce bias*
  - Avoid leading questions. Ask "What do you think is the most important issue in buying a new car?" Not, "Do you think safety is the most important issue in buying a new car?"
  - Avoid name dropping. Ask about the hiring freeze, not the new CEO's hiring freeze.
  - Avoid suggestive language. Colorful adjectives, like "dangerous" and "charming," as well as inflammatory verbs like "prohibit" and "cater," can change people's answers.
  - Check for problems with question order. Asking about Securities and Exchange Commission (SEC) regulations immediately after asking about insider trading may get you a different answer than asking about it after asking about the effects of deregulation.
- *Increase information*
  - Ask for opinions instead of facts. When asking a respondent about him or herself, phrasings such as, "Do you consider yourself a ..." may be better than, "Are you a ..." because a person may not be sure how they fit in various social categories.
  - Use open follow-ups for fixed-response questions. Fixed-response questions are easier to tabulate, but always leave the respondent room to expand, clarify, explain, or comment. They will feel better and the information may be useful.

After our questions are written, it is important to pre-test them with a small sample of our population to make sure the questionnaire works. We should also test tabulation and encoding of our sample survey. Many small companies run surveys and only realize after the fact that they'll never have time to evaluate all different answers to their open-ended questions.

Conducting the Survey

Once the plan and the budget are in place, and the materials are prepared, the next step is to actually conduct the survey. We may do this ourselves, or hire a service provider to do it for us. Here are the steps:

- *Train the interviewers.* Even professional interviewers need an introduction to the job.
- *Pick the starting time,* selecting a time that will reduce bias.
- *Collect the data* and protect confidentiality while you do it.
- *Follow-up with incentives* to increase sample size.

Interpreting and Reporting the Results

For a well-planned study, where a statistician has been brought in early, the interpretation and reporting of results can go smoothly and easily. For the poorly planned study, this is the point where disaster is likely to strike.

- *Code and tabulate the data.* Except for open-ended questions, coding was built into the survey. Issues related to coding and tabulation were discussed in Chapters 2, 4, and 6.

- *Analyze the data.* Be sure to determine the margin for error, that is, the point where a difference in value is not meaningful.

- *Evaluate unexpected results.* People are less predictable than most other subjects of statistical study. When we get unexpected results, we should consider further study, a special report, or action.

WRITE THE REPORT

The guidelines for report writing are in Chapter 10 "Reporting the Results." An important additional issue is privacy. The report must not allow a reader to identify any respondents.

Quiz

As a gift to our readers—a bonus. You have only five questions to answer—short chapter, short quiz!

1. What is the purpose of a survey?
    (a)   To be mathematically correct
    (b)   To please the boss
    (c)   To answer the question
    (d)   None of the above

2. What is the advantage of a Mail Out survey?
    (a)   Lowest cost
    (b)   Respondent convenience
    (c)   Interviewer bias
    (d)   Postal costs

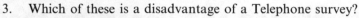

3. Which of these is a disadvantage of a Telephone survey?
    (a) Speed of response
    (b) Large numbers
    (c) Accuracy
    (d) Cell phones

4. A true or false question is a _____ question.
    (a) Multiple-choice
    (b) Dichotomous
    (c) Completion
    (d) Open-ended

5. When writing a good survey question, you should...
    (a) Reduce error
    (b) Reduce bias
    (c) Reduce both (a) and (b)
    (d) Increase both (a) and (b)

Forecasting

Specialists in finance, marketing, and economics all claim to be able to predict the future and are prepared to charge us for their services. Of course, if they can really predict the future, why can't they just make a few million on the stock market and give us their services for free? More seriously, the complexities of forecasting are intimidating. We need to learn how to be intelligent consumers of statistical forecasts.

FUN FACTS

In his classic book, *A Random Walk Down Wall Street*, Malkiel (1973) argues both that random portfolios will perform as well as those selected by experts and that short-term predictions of the stock market are impossible. In forecasting, long-term forecasts are less reliable than short-term forecasts.

The Future Is Good To Know

Forecasting is the product of two opposing forces. Nothing could be more valuable in making business decisions than knowing certain facts about the future course of our business. However, no amount of statistics or mathematics or computer power can change the simple fact that the future can always surprise us.

THE PAST IS PROLOGUE: OR MAYBE NOT

The most important thing to realize in using statistics for forecasting is that past numbers are not the cause of future numbers. Yesterday's numbers were created by yesterday's actions and events. Tomorrow's numbers will be caused by tomorrow's actions and events. When we use statistics for forecasting, we are trying to use the numbers as representative of past actions and events, and then project the numbers forward, implying a certain set of future events that we believe will happen, leading to the future numbers in our forecast. However, if future actions and events are different from past actions and events, then future numbers will not be predicted by past numbers.

TIPS ON TERMS

Planning estimates. Predictions where the probabilities of various relevant events, usually provided by experts, are coordinated within a mathematical system. We will see how this is done in the final section of this chapter. (Remember that this use of the term, estimate, has absolutely nothing to do with the statistical estimates discussed earlier in *Business Statistics Demystified*.)

Forecasting projections. Predictions derived from one of a number of statistical procedures that take a series of data measurements drawn at equal intervals over time and extrapolate from that series into the future. The remainder of this chapter will focus on these techniques.

CROSS-SECTIONAL AND TIME-SERIES DATA

In terms of statistics, we can think of this in terms of the structure of a regression model. At regular time intervals, we take a number of individual measures, some of which are important to our business decisions. In a cross-sectional approach, we regress the important variables on the other variables, either measured at the same time, or with the other variables measured at some earlier time period. In a time-series approach, we regress the important variables (or an aggregate combining them) on the time variable measuring when the measurements were taken. The goal is always to be able to predict the values of these important variables for a time period in the near future.

The cross-sectional approach can be handled by a regression analysis. The problems are as discussed above. If we regress the variables of interest on other variables measured at the same time, we will not really be looking at the future. The values for the independent variables will not be available before the time when we can measure the dependent variables for ourselves. What we need is a set of independent variables measured at one time that predicts the values of dependent variables at some later time. In economic forecasting, these sorts of predictive independent variables are called leading indicators. If such indicators can be found, the rest of the problem is just a matter of regression.

The time-series approach is more complex. As we will see below, although it is structured like a regression problem, certain vital assumptions are violated, and regression per se cannot be used. In this chapter we will consider time-series analysis, in terms of regression-like and other types of methods.

The Measurement Model

The first step in understanding time-series analysis is to understand what a time series is and the basic model of how it can be analyzed.

WHAT IS A TIME SERIES?

In a time series, we measure the same variables over and over again at equal time intervals. In a business context, this could be every day, or every week, month, quarter, or year. The variables, of course, are variables whose values, if we knew them in advance, would be helpful in making our business decisions. Each set of values is recorded, labeled with the time when the measurement was taken. A forecast involves using the entire set of values collected up to the present time to predict the next few values, sometimes just the next value, for the next few time periods.

In some cases, we will want to examine individual variables. In other cases, we will want to combine variables that measure similar things into one variable, called an aggregate or an index. The U.S. Government calculates and publishes many such indices, which measure everything from the prices of different types of oil, the amounts of various products and services sold, to the interest rates one bank charges another for borrowing money. The advantage in creating an aggregate index to analyze is that it provides a single dependent variable, which as in regression and ANOVA, simplifies the technical problems considerably.

THE CLASSICAL MULTIPLICATIVE MODEL

In considering how to analyze a time-series variable, a general model, called the *classical multiplicative time-series model*, is used. The equation for that model is:

$$Y_t = T_t \times C_t \times S_t \times I_t \qquad (16\text{-}1)$$

where Y_t, T_t, C_t, S_t, and I_t are the values of the dependent variable, the *trend*, the *cyclical*, the *seasonal*, and the *irregular* components of the time-series. The subscript, t, indicates a separate value for each time period in which data are collected.

The idea is that, at each point in time, the overall *y*-value that we want to predict is composed of four different numbers. The first number, the trend, either rises or falls (but not both) according to some specific rule, over the length of time being measured. The cyclical component rises and then falls and then rises again slowly over time. The seasonal component also periodically rises and falls, but does so in synchrony with the time of year. The irregular component consists of non-repeating ups and downs due to events not involved in any of the other components.

Note that the equation for the multiplicative model does not include an explicit error component. The irregular component is random in the sense that it is not predictable from the time variable, but it is not due to error. Each of the four components has its own equation (at least in theory) and each component equation includes the error variance for that component separately.

HANDY HINTS

Why Does the Time-Series Model Use Multiplication?

In previous chapters, whenever we combined separate elements in an equation, we almost always used addition rather than multiplication. Why does the classical time-series model use multiplication instead? The reason is mathematical convenience. For certain types of time-series analysis, it is easier to extract out the different components if we assume that they are combined using multiplication, rather than addition. For other types of analysis, it is assumed that addition is used. For those types, another model, the additive model, which is identical to the multiplicative model except for using addition instead of multiplication, is used. The two models can even be converted back and forth mathematically, if need be. The only difference is the scale of measurement. In the real world, influences combine to create an effect. The choice of which type of mathematics to use is ours.

We discuss each component below:

Trends: patterns of growth

The trend is defined as the systematic change in the dependent variable (that is, the variable of interest to us) that does not cycle over time. As such, it is a pattern either of growth or decay over time. The trend is either up or down, but not both. In product sales, it represents the increase in sales from introduction until saturation. In terms of personnel, it represents the number of employees from the company's start until it reaches its full size. The different types of trends correspond to the different patterns of fast and slow change over time.

Most forecasting is about how fast, and in what way, something will grow. We can identify four basic growth patterns, and then discuss cycles.

- *Flat growth* is no growth at all. This is what we will see in the salesman's results in the story below. Year after year, $5,000 per month. Mathematically, a horizontal line, with a slope of zero.
- *Linear growth* is a steady rate of growth, adding a fixed amount per month or per year. Mathematically, a sloped straight line.
- *Exponential growth* is accelerating growth. It can be a slow acceleration, or a fast one. Mathematically, a positively accelerating exponential curve.
- *S-curve growth* is growth according to a more complex model. Here, we have a slow beginning, a rapid increase, and then a leveling off. The mathematics is complex.

SOMETHING EXTRA

When Does the S-curve Curve?
The S-curve model is used because it matches the observed reality of many systems in business and in nature, such as: Slow startup, rapid growth of sales until the market is saturated or matures, and then sales level off; and slow increase of size from seed to sprout, followed by rapid growth of a plant, followed by slow growth as the plant reaches maturity. What creates this common pattern? The slow initial growth is due to internal constraints: in a business, not enough money or employees; in a plant, not

enough leaves and roots. The rapid period is when, due to exponential initial growth, the internal limits are released, and no external limits are yet met. This period ends when external limits—such as saturation or complete use of local resources—are reached, or when new internal limits—such as too many communications channels—arise.

The key to S-curve forecasting, then, is to know what limits growth, and when these limits will hit. Statistics can help with this, but we also use modeling and expert judgment.

Let's take a look at each of these five types of growth in Table 16-1. We will start with an investment of $100. The first column shows what you will have over ten years if you hide your money in your mattress and earn nothing—there is no growth. The second column is an example of linear growth—a simple bank account at 5% interest, but you remove the interest every year and stuff it in your mattress. The third column shows slowly accelerating growth achieved by leaving the interest in the bank account. The fourth is the result of a very successful—or lucky—entrepreneur who invests in a series of venture capital deals with a 30% annual return, and all the businesses succeed. Such exponential success is unlikely. The fifth column shows the S-curve, a more realistic example. It might show the results of a savvy entrepreneur who invests her own money in her own new company. At first, while she is getting the business started, she gets very little money back. Then business takes off and the money rolls in. Finally, business reaches saturation and there is slow growth. Soon, it is time to retire. Note that the same proportions would arise with more realistic investment numbers in the tens or hundreds of thousands of dollars. Note also that each pattern of growth is also a pattern of shrinkage if the direction of the trend is reversed.

In these examples it was fairly easy to forecast how much money we would have in each future year. Why? Because we knew the model; that is, we knew the rules. We clearly understand the consequences of each action we take—at least for the first three columns. With venture capital and investing in our own company, there are no guarantees. We are hoping and guessing.

Predicting business growth or stock market results is much more complex. And therein lies the danger. If we have a desire—such as a political motivation—to show a certain result, we can always find assumptions that will lead to that result. The challenge in forecasting—and in evaluating the forecasts of others—is not so much the application of the statistics as it is the choice of underlying assumptions. We must also be aware of tunnel vision. The moment we say, "given these assumptions," we should take time and energy to ask, "Should we accept these assumptions? Should we accept that they will continue to be true in the future?"

Table 16-1 Five types of growth.

	Flat	Linear	Slow acceleration	Fast acceleration	Life cycle
Investment	Mattress	Bank account, remove interest	Bank account, retain interest	Venture capital	Own company
Initial investment	$100.00	$100.00	$100.00	$100.00	$100
End of year					
1	$100.00	$105.00	$105.00	$130.00	$105.00
2	$100.00	$110.00	$110.25	$169.00	$110.25
3	$100.00	$115.00	$115.76	$219.70	$115.76
4	$100.00	$120.00	$121.55	$285.61	$137.99
5	$100.00	$125.00	$127.63	$371.29	$166.26
6	$100.00	$130.00	$134.01	$482.68	$202.36
7	$100.00	$135.00	$140.71	$627.49	$248.59
8	$100.00	$140.00	$147.75	$815.73	$307.98
9	$100.00	$145.00	$155.13	$1,060.45	$323.38
10	$100.00	$150.00	$162.89	$1,378.58	$339.54
Growth over 10 years	**0%**	**50%**	**63%**	**1279%**	**240%**

CRITICAL CAUTION

Assumptions, Assurances, and Control

From a business perspective, there are three critical issues in evaluating the reliability of a forecast. The first is the assumptions. The business assumptions in a forecast include the statistical assumptions, and a whole lot more. They include assumptions about the accuracy of data and about the behavior of people and businesses. For example, the predictions of continued rapid growth in the U.S. energy industry around the year 2000 were based on two assumptions that turned out to be false: that earnings were being reported accurately, and that the markets were not being manipulated. Due to the alleged activities of Enron and other companies, these assumptions were not true, and the forecasts turned out to be woefully inaccurate.

An assurance is a special case of an assumption. Interest on a bank account is a good example. We have more confidence in our assumption because banks are regulated and monitored, and because our investment is insured (up to a point) by the U.S. government. Regulation, monitoring, and insurance increase our assurance that an assumption will continue to hold true in the future.

Lastly, there is the issue of control. Forecasts of what our own company will do are—if carefully prepared—inherently more reliable than forecasts of what other businesses, the general market, or consumers will do. Why? Because, as executives, we can direct our company to engage in the activities we have determined will bring about forecasted desirable results. To the extent that we have effective control, we can work to make our forecasts of our future happen.

Cycles

By definition, the cyclic component of the multiplicative model is a measure of a cycle that repeats over the course of more than a year's time. So, ordinarily, if our sales regularly rise and fall over the course of the week, those short-term components will not be counted as part of the cycle. Cycles usually repeat every 2–10 years. It is important to have data collected over a long enough time period to capture the cyclic component in our statistical analysis.

A major cycle in most forecasting models is the cycle of the overall economy from boom to bust and back again. Almost any aspect of our business we want to measure over time will be affected to some degree by the business cycle, and that is why a cyclic component is included as part of the model. Our overall sales may be increasing as our company grows, but, if the economy starts its downturn next month, next month's sales may be lower than this month's despite our growth. This is a good example of how the time-series model works. Each component is thought of as one thing influencing the overall value. We put all the influences together and we can predict the value for the next time period.

Seasonal

Many business variables of interest to us are affected by the seasons. Retail sales of gardening equipment sell more in the spring. Different grades of oil are used in cars during the winter than in the summer. If we are predicting values of these variables for periods less than a year, we will need to adjust for seasonal variations, and that is what the seasonal component of the model allows us to do.

Irregular

A new competitor enters the market. A new trade agreement lowers the price of a key ingredient. A snowstorm increases sales of shovels. Individual events that are not part of any pattern over time are modeled with the final component, called the irregular component. Since these events are not related to time in a reliable way, we usually measure the irregular component by extracting the other components and seeing what is left.

Descriptive Statistics

In time-series analysis, there are two principal uses for descriptive statistics. We can combine, or *aggregate* our dependent measures into one single variable that measures what we want to forecast. We can alter, or *smooth* the shape of the curve across time to make the patterns easier to see (and analyze).

AGGREGATE INDICES

Before we can make forecasts, we need to be able to graph out our data as a time series. This requires that we have a single variable that measures the value we want to predict to aid in our business decisions. Each forecast involves predicting the values for just a single variable only one or two, perhaps three, time periods out into the future. It is often a good idea to limit the number of forecasts we make to just a few critical numbers, unless we have a way of combining the forecasts for each separate variable in order to help make our business decision. (Some planning systems combine things in this way.)

If we are interested in our overall sales, we need to aggregate sales of all our products in all our sales territories. If we are interested in food prices, we need to aggregate prices of different foods. If we are interested in demand for our products, we need to aggregate sales of all products of that type, or else get ahold of the numbers from someone else who has aggregated them for us. Governments, NGOs, and private firms all publish various indices that may be useful in our forecasts.

There are two basic approaches to aggregation. We can just add up the different related variables for each time period. Or we can weight the different variables according to some measure of their importance. For example, depending on what we will use it for, we might not want to count the price of bread the same as the price of caviar in our index of food prices. Not only is bread a lot less expensive than caviar, but lots more bread is sold, the proportion of food budgets spent on bread is much higher, a larger proportion of the buying public buys bread, bread is purchased more regularly, bread comes in a lot more varieties, etc. There are many ways to weight the different variables.

Unweighted indices involve a choice of which variables to include. Weighted indices also involve a choice of how to do the weighting. In either case, there are many, many ways to build each index. The advantage to designing our own index is that we can customize the formula to best help make our business decisions. The advantage to using a standard, published formula or index is that our analyses can be compared to those done by other organizations, which is often valuable. The disadvantage is that, over time, we are locked into whatever changes the builders of that index make to adjust their formula.

SMOOTHING: FINDING TRENDS

After we have selected or calculated our dependent variable, we graph it on the y-axis, with time as the x-axis. There are simple ways to modify these graphs to make the various regular components more visible for inspection, by ourselves or the decision makers. These measures are called *smoothing functions*. The two most common smoothing techniques are the *moving average* and *exponential smoothing*. Both of these techniques average the values of the variable over time, which tends to eliminate individual changes that occur only once. Smoothing always tends to erase the irregular component. Some smoothing, correctly applied, also eliminates cyclic or seasonal components as well.

A moving average involves taking the average of all of the values for a variable over a fixed number of time periods and graphing the data point at the place on the x-axis corresponding to the middle of that time range. If our data are collected more often than once a year and the moving average is calculated for a whole year around each data point, we can use the *ratio-to-moving-average* method to *deseasonalize* the data, so we can see the *seasonally adjusted* values. In terms of the multiplicative model, the moving average for each time point includes only the trend, T_t, and cyclic, C_t, components. When we take a ratio, dividing the data values by the moving average, that leaves us with a time-series that, in principal, only contains the seasonal, S_t, and irregular, I_t, components. Then we average the ratio over all the years of the time-series (averaging out the irregular component). This final number is called the *seasonal index*. We divide the original data by the seasonal index and get a look at our data without the seasonal fluctuations. Seasonal indices for many industries and markets are also published by the government and can be used with the same advantages and disadvantages of other published indices.

Exponential smoothing takes an average extending backward in time, weighting the most recent data points most heavily. The idea is that, while the future may be like the past, the near future will be more like the most recent past than the distant past. The exponentially smoothed time series allows us to predict the very next value of time series based on the available data, but taking the most recent data more seriously. In a situation where irregular influences tend to last longer than one time period, the exponential method can be very useful.

Inferential Statistics

There are many methods for making inferences about time-series to predict the next few values of the dependent measure for a few time periods into the future. Most of these are very technical and complex. The simplest use the smoothing techniques discussed above. Each smoothing formula involves accumulating data from other time periods. For each smoothing formula, a separate formula for calculating the next value in the time-series can be constructed. These sorts of inferences can be used only to predict one or two future values.

The next more complex type of inferential method are those based on least-squares measures of residuals, the same sort of mathematics used in regression. We discuss these and other more advanced techniques below. The most important issue in using any sort of inferential technique for time-series analysis is to validate the method against our data. Because time-series forecasts always involve extending a time-series past its original range, there is a very simple method for validating even the most complex inferential technique. Just pick a time in the past, say four-fifths of the way through the time-series. Apply the inferential method(s) to the first four-fifths of the data and match the forecasts created to the last fifth of the data. In seeing how well our forecasting model has performed on recent data, we can assess its ability to perform in the next time period or periods.

LEAST SQUARES TECHNIQUES

Recall that the regression lines are just the unique line that passes through our data with the minimum of the sum of squares of the length of all the residuals. This measure is called the least-squares. We can perform this same calculation on our time-series, with our dependent variable, Y, and the independent variable, time, t. If our trend is linear, the least-squares line may capture it nicely. Even if our trend is non-linear, the least-squares line will show us the direction of the trend. We can also use non-linear regression calculations to detect non-linear trends in the least-square curve.

Using the least-squares line (or curve) will be slightly different from most regression in that there will be only one data point for every value of the independent variable. More importantly, this analysis will violate a critical assumption of regression. Because each data point is a measure of the same variable taken from the same individuals, but at a different time, we can expect that the values will not be independent. This makes sense in terms of the notion of independence and information. One of the best sources of information about next month's sales is this month's sales.

Forecasting is extrapolation

Note that, in regression terms, any inferential prediction from time-series data is an extrapolation. Our data are about the past. The range of the independent variable, time, stops at today. Forecasting means predicting Y-values for the future, outside the range of the data. All of the problems of extrapolation apply.

The problem of autocorrelation

More serious than the problem of extrapolation is the problem of the non-independence of the data points. Various inferential techniques attempt to help correct this problem by eliminating the influence of earlier data points on each of the later data points. The effect of the influence of earlier data points on later data points is called *autocorrelation*. There are a number of different ways of measuring autocorrelation. We can examine these measures graphically to help decide whether a particular inferential technique is appropriate. Some inferential techniques automatically incorporate certain measures of autocorrelation into the inferential process.

Choosing among models

Even if we just restrict ourselves to least-squares techniques, we have many, many options. There are linear, quadratic, cubic, and exponential functions just for a start. An expert in forecasting is the most important thing to have in choosing among models. Computers and statistical software make it much too easy for the non-expert to perform forecasts using one or many different techniques. The computer gives us lots of useful numbers, but it can't provide the experience and expertise to use them effectively.

There are not only many, many inferential techniques, there are also many ways of comparing how well different techniques work. These comparisons help us choose which techniques to use in making our forecasts. The basic strategy in comparing techniques is to run each technique against our data and see how well the predicted values match the actual values over the range of past times. The match is calculated by combining the residuals and is called a *fit*.

In general, more complex inferential techniques involving more parameters will give us a closer fit to the data. Choosing a model means picking the one that fits the data best, given the number of parameters. This is a complex process, involving judgment and a comprehensive understanding of the details of the differences between the models being compared. Each step in the process can be assisted by the computer, but each step involves more and more numbers to understand and evaluate.

ADVANCED TECHNIQUES

Some advanced inferential techniques use more complex smoothing techniques. Some, like the Holt-Winters technique, combine the prediction based on the smoothing technique with other predictions. (In the case of Holt-Winters, the smoothing is combined with a trend prediction.) Many advanced techniques use more advanced types of least-squares techniques. The most common of these attempt to fit a non-linear trend to the data. When the shape of the curve gets sufficiently complex (for instance, an S-shaped curve), least-square calculations cannot be used to get a solution. An alternative calculation technique, called maximum likelihood, is used to fit the data to these more complex trends.

Maximum likelihood methods used to match S-shaped trends to data require a specific mathematical form of the S-shaped curve. The two most common mathematical functions used are the logistic function and the Gompertz function. Maximum likelihood methods can also match data to much more complex curves. For example, when irregular changes affect the shape of the time-series curve, different functions may provide the best fit for different segments of the time range. If we can identify the times when these changes occur, the use of splines may be effective.

As mentioned above, there are techniques called autoregressive (or AR) techniques that use an underlying model of autoregression to predict future values in the time series. The most common of these is the Box–Jenkins technique. The underlying model is that each Y-value is predicted from the preceding Y-values instead of from the time. AR methods do this by taking into account the autocorrelation, which measures the correlation between earlier and later values. There is also a sophisticated forecasting method called ARMA (or ARIMA), which combines the autoregressive and moving average techniques.

There is also an inferential technique based directly on the classical model. Spectral analysis uses the mathematics of Fourier transforms to decompose the time series into a set of cyclical curves. This is a very complex technique. Its use and interpretation requires an expert.

In recent years, various types of mathematical modeling, including neural networks (which are equivalent to a form of non-linear regression and can also model autocorrelation using additional calculations called recursive connections) and genetic algorithms (a non-statistical technique originally developed for modeling biological systems), have become available for forecasting, especially forecasting stock market prices. The jury is still out, but some of these techniques have been shown to work well for specific kinds of data for periods of time.

Cautions About Forecasting

Using statistics or other mathematical or computational techniques to attempt to predict the future, even if only in the limited terms of a stock price or next quarter's profits, is somewhat more reliable than using tea leaves or a crystal ball. However, there are essentially insoluble problems faced by forecasters that are not faced for any of the statistical techniques we have discussed earlier in *Business Statistics Demystified*. A crucial aspect of understanding forecasting will be to understand the limitations of its methods. We divide the discussion of these limitations into problems of the methods and problems of the world.

METHODOLOGICAL LIMITATIONS

We have already discussed the problems of extrapolation, single values at each time point, and non-independent, autocorrelated measures. These problems seriously limit what we can prove about the effectiveness of forecasting techniques. However, there are inherent limitations to the methods used that are not due to these considerations.

Graphics versus calculations

The simplest techniques for time-series analysis rely on graphics. We change the data values and look to see the new curve and attempt to interpret it. However, extraordinarily complex mathematical techniques used in time-series analysis sometimes have a step where mathematics is abandoned and human judgment, assisted by a graph, must be employed. This is particularly true when different models are compared for fit.

The problem with this is that, as psychologists have long known, people see what they want to see. If we want our market share to be increasing, we are more likely to see an increasing trend even when one is not there. (Psychologists call this experimental bias.) This phenomenon is far more pervasive and far more powerful than most folks imagine. Psychologists have demonstrated that even when the person making the judgment has no bias, just being in the room with an experimenter who is biased can affect the results! At a very minimum, we must be very cautious about both expert and non-expert judgments. No amount of complicated mathematics will eliminate the effects of a biased judgment applied somewhere in the process. If possible, we should have disinterested third parties double-check our assessment of trends and be sure not to give them any hint of what we are looking for. However, this is difficult, time-consuming, and expensive.

Conservatism

Notice that we have not talked about α-levels or Type I errors here. Because of all of the problems with predicting the future, there is no way to put an upper limit on the probability of our forecasts being wrong. The inferences made in time-series analysis are not conservative. We are given no guarantee that our forecasting techniques will give accurate predictions, even if those techniques have produced accurate predictions in the past. The only guarantee we have is that predictions for events further in the future will be less reliable than predictions of events sooner to come.

Correlation of variables and results

Underneath every forecast is a model of how reality works. That model may be wrong. It is best to know what our model is.

An old story illustrates this point well. There was a salesman who, starting as a young man, was given a small territory. He generated $5,000 a month in sales. That was spectacular for such a small territory. The business evaluated his sales dollars per customer. It was really high, so they thought he was a great salesman. They gave him a bigger territory, expecting great things. In the new territory, he generated $5,000 a month in sales. That was reasonably good, measured in sales dollars per customer. And he stuck with it. Eventually, due to seniority, he was given a bigger territory. He generated $5,000 a month in sales. That was a really poor return, measured in sales dollars per customer. And he was getting old, too. So they gave him the smallest territory they had. He generated $5,000 a month in sales, which was pretty remarkable, measured in sales dollars per customer.

Looking back, it is easy to see that the company was correlating to the wrong variable. Dollars of sales was a function independent of number of customers. The correlation was with the salesman. He was a $5,000-a-month man. If they'd known that from the beginning, they would have kept him in the smallest territory all those years, and everyone would have been better off.

So, before we dig into the numbers, there are some questions we should ask about our model:

- Does the model have a strong theoretical foundation? Do we understand it?
- Is the model proven through long use delivering accurate results? Has it been modified or improved? Have the changes been tested?
- If the model is being borrowed from another situation, have we investigated the underlying assumptions that worked there, to see if they apply to our situation?
- What are the assumptions of the model? Have they been reviewed and evaluated recently?
- Is the model, or are certain elements of the model, being chosen in order to achieve a desired prediction? We see this every year with the different predictions about the size of the national debt. One political party uses one set of economic projections, another uses a different one. But they choose the projections to get the result they want. These are not really projections at all; they are something closer to wishful thinking or spin.

EXTRINSIC FACTORS

Nature seems to resist our attempts to know the future. This not only shows up in all of the various technical difficulties with forecasting, but also in our commonsense understanding of the world.

Error and change

Things change. As discussed above, the success of any forecasting technique depends critically on our model of the world. When the world changes, the variables that used to provide wonderful predictions may provide terrible ones and previously useless variables may become useful. The best example of this is published indices. The formulas for the various published indices of the economy, the stock market, trade, particular industries, etc., are regularly modified by the organizations that calculate and publish those numbers. The Dow Jones Industrials is a group of companies whose stock prices is intended to measure the state of the stock market and to some extent, of the economy as a whole. Only one company included amongst the Dow Jones Industrials at the stock market crash of 1929 was still included in 1945. Not only do the firms change, the relative contributions of different industries change as well. Imagine how the relative contributions to the economy of manufacturing, natural resources, energy, retail, transportation, and communications have changed over the last 100 years.

Changes that are due to disruptive events, such as wars, flood, and famine, can be modeled as part of the irregular component. But changes that alter the underlying causal structure will tend to change which model and which type of model make the best forecasts. Even if the same variables are useful in predicting the things we want to know about, changes to the way things work can alter the underlying causal relations and thus change which model fits best. For example, changes to regulations, such as SEC, FCC, or FAA regulations may not alter what we need to measure, but may radically alter the shape of the curve.

Nothing grows forever

In the wake of the burst of the Dot-Com bubble, we have to re-examine how we think about forecasts. Almost everyone predicted the bubble would burst. Corporate growth driven by investment of venture capital doesn't last, it has to be replaced by earned revenue. But why were our projections so unrealistic, and why did we believe them for so long? The authors suggest two main reasons. First, each company was predicting success of its own audacious business plan, but not looking at the big picture of the whole industry. Second, everyone was invested, not only financially, but personally. Anyone with money in the system wanted to believe their phenomenal success would keep growing. But nothing grows forever. And no prediction about when growth will stop is proven until after it happens.

Quiz

1. Predictions that take a series of data measurements drawn at equal intervals over time and extrapolate from that series into the future are known as...
   (a) Planning estimates
   (b) Forecasting projections
   (c) Both (a) and (b)
   (d) Neither (a) nor (b)

2. Combining several variables that measure similar things into a single variable will result in a _____ variable.
   (a) Aggregate
   (b) Time-series
   (c) Forecast
   (d) Regression

3. The _____ component of the *classical multiplicative time-series model* describes the pattern of either growth or decay over time.
   (a) Cyclical
   (b) Seasonal
   (c) Irregular
   (d) Trend

4. The _____ component of the *classical multiplicative time-series model* describes a pattern that repeats over the course of more than a year's time.
   (a) Cyclical
   (b) Seasonal
   (c) Irregular
   (d) Trend

5. Which of the following involve descriptive statistics?
   (a) Moving average
   (b) Exponential smoothing
   (c) Both (a) and (b)
   (d) Neither (a) nor (b)

6. The main problem with using the least squares technique is...
   (a) Extrapolation
   (b) Autocorrelation
   (c) Interpolation
   (d) Smoothing

7. In general, more complex inferential techniques involving more parameters give us a closer _____ to the data.
    (a) Comparison
    (b) Smooth
    (c) Fit
    (d) Autocorrelation

8. Most time-series analyses rely on _____, which are subject to interpretation bias.
    (a) Graphics
    (b) Experts
    (c) Non-experts
    (d) Conservatism

9. When evaluating our model, we should ask . . .
    (a) Does the model have a strong theoretical foundation?
    (b) Is the model proven through long use delivering accurate results?
    (c) What are the assumptions of the model?
    (d) All of the above

10. Which of the following is an extrinsic factor for forecasting failures?
    (a) Error
    (b) Change
    (c) Nothing grows forever
    (d) All of the above

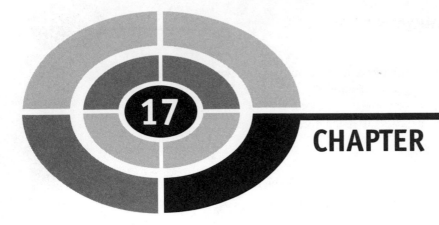

CHAPTER

17

Quality Management

Quality management is a very interesting application area for the business statistics student because we can see how statistical methods are integrated with engineering and business processes to create process changes that lead to measurable improvements. Although quality management is a very broad field, including both harder engineering applications and softer, team and people-focused techniques, the statistical processes fall on the more technical, more defined, and measurable, engineering side of the field.

Key Quality Concepts

To understand how we use statistics to improve quality, we need to see how a quality engineer looks at the world. Although there are larger, more general definitions of quality used on the softer side, the most relevant definition of quality for our purposes is: *quality* is conformance to specifications. For today, let us presume that the soft-skilled managers have done their job well, and that we have a specification, which, if met, will deliver what the customer wants. Now, it is our job to ensure that the process we manage delivers a product that conforms to that specification. This kind of quality management can apply in manufacturing environments, where the process is an assembly line and the products are mass-produced, and it can also apply in project environments where we are producing a single result—perhaps a large computer program—and we want it to be free of defects.

TIPS ON TERMS

Defect. A single instance in which one product has a single variable whose measurement does not conform to specifications.
Error. An event that leads to a defect, being its cause or part of its cause. Note that error has one meaning in statistics, and another in quality management.
Root cause. A single cause identified as being at the root of many problems. If eliminated, many defects go away.
Incident. A single event in which some cause leads to some defect.
Event. A single occurrence of a single input or process.

We can think of quality management for process control as the effort to reduce the defects from a defined process to an acceptable level. Our specification is a set of standards applied to measurable aspects of each product, usually presented as a range of acceptable values for each measurement. If we measure our product, and a variable falls outside the acceptable range, that is a defect. We also measure various aspects of our work processes. If we can correlate a value in a measurable process element with occurrence of a particular defect, that correlation may indicate that the error is a cause of the defect. If we eliminate the error in the process, we then reduce the frequency of the defect, increasing quality.

KEY POINT

There are Always Errors
A key principle of quality management is that errors are always present. We seek to reduce their number and consequence, and keep them at an acceptable level.

This kind of quality management calls for a combination of engineering and statistical methods with one unified focus: reducing the level of defects to an acceptable level and keeping it that low. We will look at two different quality management methods that use statistical tools: root cause analysis, which uses a Pareto diagram, and process control, which uses statistical sampling and measures variance.

Root Cause Analysis

Root cause analysis can be applied to any environment where we control a process—either a physical process performed by machines, or a business process with work done by people—that has measurable outcomes. Root cause analysis is a very effective method of process improvement because it identifies the relatively few causes of a large number of defects. Correcting those causes leads to a very quick reduction in the number of incidents in which defects occur. To perform a root cause analysis, we combine an engineering tool (the fishbone diagram) and a statistical technique (Pareto analysis).

The fishbone diagram, also called an Ishikawa diagram, was developed as a process improvement tool as part of the Total Quality Management (TQM) movement in its early years in Japan. As you see in Fig. 17-1, the Ishikawa diagram is a map of a repeating process where a variety of inputs are combined through a variety of processes to produce a product. Our first step would be to create such a map of our production or work environment. Our own diagram may have any number of inputs and processes, and we may define intermediate sub-components and series of processes. The goal is to have a complete picture of all the events that lead to the creation of a product.

To ensure quality, we measure all relevant features of a sample of our products. We also measure as many features as possible of a sample of the inputs and of our processes. The principle underlying the Ishikawa diagram is that a defect in the product can be traced back to one or more causes that appear as variations from the norm—or variations from internal specifications—in the inputs or processes. The engineer's goals are to define all inputs

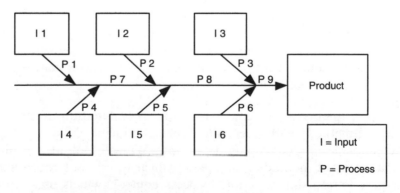

Fig. 17-1. Generic fishbone or Ishikawa diagram.

and processes, and define the relevant variables for each of these, and to specify the quality requirements for each input and process. Theoretically, if each input meets specifications and each process is within tolerances, every product is defect free on all features. In practice, iterative application of this method does produce amazingly high levels of quality that meet business requirements and, when appropriate, approach zero defects.

HANDY HINTS

Sometimes, Zero Defects is Really What We Want
The cost of creating a defect-free product is extremely high. However, sometimes it is worth the effort. For example, when we are creating a system with very tight tolerances where human life is on the line, such as microsurgical robotic devices, jet fighters, or the space shuttle, we aim for zero defects. In such systems, any defect or combination of defects could lead to a catastrophic failure, which then leads to the death of people. Zero-defect initiatives were designed to support such projects. In other situations, some degree of risk is acceptable. One alternative to zero defects is redundancy. For example, elevators have three separate braking systems. Any one of them might fail. But, given regular inspection, the odds that all three will fail simultaneously and the elevator will plunge are extremely low.

In the data we gather, we will have a series of product defects, $D1$, $D2$, $D3$, ... We also have a set of potential causes, defined as measured results where our data about inputs and processes indicate a variable outside its specified tolerance. Let's call these causes $C1$, $C2$, $C3$, ... Each defect has one or more causes, and we can identify those through a combination of statistical and engineering analysis where a particular cause $C(x)$ will be linked to a particular defect $D(x)$. This linkage defines an incident, $I(x)$, where we can say that some cause or set of causes resulted in a particular defect.

Over a period of time, different incidents will repeat with greater or lesser frequency. If we plot the frequency of incidents, by type of incident (a nominal variable), we find that the distribution of incidents has a particular pattern described by Pareto's law. The derivation of Pareto's law is beyond the scope of this text, but it has a solid foundation in both theory and tested practice. Pareto's law—also called the 80/20 rule—states that 80% of all defects arise from only 20% of the underlying root causes. In order to prioritize our quality improvement work, we generate a Pareto diagram, a bar chart with a nominal variable on the X-axis.

Let us look at an example of how this works. This example is based loosely on a consulting contract performed by the second author of this text. It is an interesting case because it shows how root cause analysis, originally developed for improvement of automated manufacturing, can be applied to a more loosely defined work environment with people engaged in many different activities. A large agency had a computer support desk supporting over 3,000 computer users. The agency was in the middle of the rollout of new workstations for every user. For a variety of technical and administrative reasons, the rollout was not going well, and there were many defects in the new operating system and computer system that users were getting. When users had trouble, they called the help desk.

Our population of incidents was all events when a user experiences difficulty completing work on one of the new workstations. Our sample—large, but probably biased—was the incidents that were called in to the help desk. The help desk call log provided our raw data. We cleaned up our data by analyzing calls and consolidating multiple calls from one person about one problem into a single incident.

For each incident, the help desk sought to identify the activity the user was trying to perform, the applications in use and running, and the symptom seen by the user. These problems defined by the help desk went to the engineering staff as their input to root cause analysis.

The engineering team—led by the secondary author—engaged in root cause analysis. We had the help desk's list of problems. We had technicians going out to fix problems and submitting reports of the causes discovered, and how the problem was resolved. We had engineers review the components and processes used in assembling the computers and for installing each software component. We had managers contact vendors and send samples of the most difficult or urgent problems to get the vendor's information about whether their input might be the cause of the problem.

SOMETHING EXTRA

Finding and Remedying a Root Cause

In this type of process improvement, it is important to identify root causes, to distinguish them from intermediate causes, and to perform an action that permanently eliminates the root cause and all resulting problems. For example, customers reported two problems: complete failure of the machine to start, and complete lockup of the machine, requiring a power-off restart. Individual technicians reported many of these incidents as being caused by the problem "memory module failure," and the solution as "replaced all memory modules." That one computer was now fixed. However, the engineers were able to trace almost all of these incidents to one brand of machine in which the memory modules did not meet the specification required by the operating system. We took a proactive approach, replacing all the memory modules on all of the computers of that brand, whether or not failures had been reported. And, due to a well-written contract with the vendor, we were able to initiate action that led to the vendor paying for the repairs. As a result, all workstations met the operating system specification, and we were confident that we could say that, throughout the organization, this incident would not happen again. When a solution to a root cause is applied across an organization, that is called permanent preventative solution. Note that a permanent preventative solution has the result that incidents of that type drop to nearly zero. We are not seeking to manage incidents better when they happen, we are seeking to eliminate incidents of error altogether.

At the peak of the crisis, let us say that the help desk was experiencing an average of 550 calls per day. Our root cause analysis identified 22 different root causes that, together, explained all of these incidents. Our full Pareto diagram has 22 bars, as illustrated in Fig. 17-2. (Note that Pareto diagram and Pareto chart both mean the same thing.)

Each type of incident is a single defect attributed to a particular cause or set of causes. A partial list of defects might include:

- Failure to boot: brand X computer with bad memory modules
- Failure to boot: computer with mixed brand of memory modules
- Failure to convert word processor document: incompatible printer type
- Failure to convert word processor document: embedded graphic conversion failure

It may seem odd that we can apply Pareto's law, which is based on a statistical distribution, when our list of incidents is a list of cause–effect pairs.

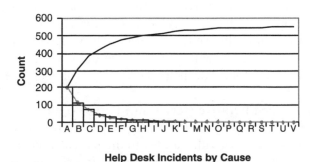

Fig. 17-2.   Pareto chart: computer help desk incidents by cause.

We are able to because, while there is a cause–effect relationship within each incident, the incidents are independent of one another. One incident does not cause another, and therefore they are independent. As a result, their frequency falls on a curve determined by Pareto's law.

This helps our help desk. Here's how. Prior to this effort, the help desk was snowed under and trying to handle each call individually. A team of fewer than fifty people can't handle 550 calls per day when all of them involve new equipment, and many involve replacing parts, reinstalling software, or contacting vendors who may take days to provide support.

The Pareto diagram cuts to the root and defines the most important work to be done. In Fig. 17-2, we have ordered the incident types from highest frequency of occurrence to lowest. The descending gray line—and the bars—show frequency of each incident, while the rising line shows the cumulative frequency. We now see that if we can fix the first four or five problems (20% of 22 problems), we will eliminate 456 out of 550 (or 82%) of incidents. That's very close to the Pareto's law prediction of 80%. Fixing these problems means going from 550 calls per day to 99 calls per day. The engineering team did just that.

The result was a whole new environment. Productivity increased in the agency. Customers were more willing to call the help desk because they saw problems being solved. As a result, our sample became more complete. In some cases, new problems were found. Most of this was due to one problem masking another. When my computer won't start, I can't convert any documents. Now that my computer starts, I find that I can't convert documents, and I place another call.

The engineering team generates a Pareto diagram for the new environment. This is much easier, because most incident types have been defined, although new ones may always arise. Due to hidden and previously unreported errors, we find that, instead of 99 calls per day (18% of 550), we are at 125 calls per day, from 17 different causes. Performing permanent preventative solutions on the three or four most common problems should reduce that to about 25 calls per day. This is a manageable number for the help desk, and too small a figure for Pareto analysis. If we want to continue to improve performance, we will now analyze data weekly, then monthly, then annually. Pretty soon, our help desk staff will feel like the Maytag repairman.

This process is called iterative (repeating) Pareto optimization. When it is applied in a rapidly changing environment, or an environment where many inputs and processes are not fully under management control, it leads to a low level of incidents—a manageable amount of interference with work, and a manageable load for the help desk—in an environment where new causes of problems constantly arise. We also find that, as we proceed down this road, some causes are extremely expensive to fix. For example, once all hardware and software technical problems are resolved to a manageable level, most incidents will arise from a need for user training, which can be costly.

If Pareto optimization is applied in a stable, controlled environment, such as automated manufacturing, extremely low error levels can be reached very quickly. This brings us to the realm of quality control for process control, our next topic.

Statistical Process Control

We now move to a manufacturing environment that produces millions of products—say computer chips or flat screens for TVs or computers. Each product contains millions of components. (For example, a single 512-megabyte computer memory module has over 36 million transistors, and if one of them doesn't work, the module won't work.) Or, we could look at automobile manufacturing, where we make hundreds of thousands of cars with tens of thousands of components. In either case, each input and process is highly specified. In fact, our company has probably been working with its vendors for decades, sharing process improvement methods to reduce the frequency of defects in our inputs. So, our Ishikawa diagrams and Pareto charts are already in place.

In this kind of environment, engineers define quality goals in terms of statistical variance or sigma (σ). Through a variety of processes, including Pareto optimization, Total Quality Management (TQM) brought many U.S. manufacturing environments—steel production, automotive assembly, and electronics, among others—to the 3-sigma level—that is, to fewer than 27 defects per 10,000 events—from the 1970s to the early 1990s. The exact conversions from statistical terms of 3-sigma and 6-sigma to the same terms as used in the quality management field are complicated by two factors. One is that sometimes we want to eliminate extreme variations on both sides of the mean, and other times, we are concerned about the range entirely on one side of the mean. The other is that there is an observed phenomenon that, given a sigma value over a short period of time in a manufacturing environment, the sigma value over a longer period for that environment will be 1.5 sigma less. So that a measured short-term of 6 sigma is expected to give a long-term result of only 4.5 sigma. Given these interpretations, the derivation of sigma value is beyond the scope of this text. Table 17-1 shows generally accepted sigma values.

Table 17-1 Industry accepted sigma (σ) values.

Sigma	Nondefects per million opportunities	Defects per million opportunities	Nondefect percent	Defect percent
3	993,200	66800	93.32%	6.68%
4	993,790	6210	99.379%	0.621%
5	999,680	320	99.968%	0.032%
6	999,996.6	3.4	99.99966%	0.00034%

An opportunity for defect can be seen as either an event in a process, or as a component in a product. The goal is to have the target number of nondefects per million opportunities, that is, the target number of results within acceptable quality specifications, for each measured variable. Of course, a great deal of engineering goes into deciding what should be measured, setting acceptable limits, and discovering interactions among processes and their consequences.

In the 1990s, various companies then raised the bar to 4-sigma, 5-sigma, and 6-sigma, or fewer than 3.4 defects per million events. Six sigma became an industry buzzword after GE announced an intention to reach this level in all its business areas in 1996. In reality, some industries—such as computer chip manufacture—need and achieve far higher levels of quality, and others find moving from 4- or 5-sigma to 6-sigma costs more than it is worth. Manufacturing quality goals should be set by calculated return on our investment of quality, or a cost of quality study, which is beyond the scope of this text.

One commonly used tool for improving quality is statistical process control, which is illustrated in Fig. 17-3. Here, we are looking at variation from the mean of a single variable in our product with samples taken over time.

As samples are taken over time, they are plotted on the graph. Sample values are shown as circles in Fig. 17-3. The mean value for the variable is targeted to be the optimal value for the product. It is centered between the specification limits shown as lines at the top and bottom of the diagram. For example, if we are producing ball bearings, we may aim to produce them with a mean diameter of 1 mm, and a specified limit of variance of +/− 0.01 mm. In the ideal case where there are no errors in our manufacturing process, the diameter of our ball bearings will vary in a normal distribution around the mean due to random variations in the process, called chance or common causes. Interestingly, if we have lots of problems, the diameter of our ball

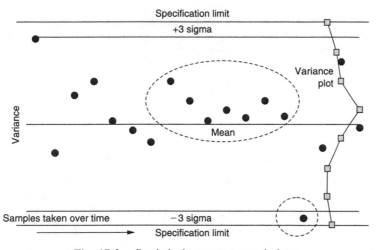

Fig. 17-3. Statistical process control chart.

bearings will also vary around the mean in a normal distribution. Each problem will tend to make the bearings too large, or too small, by a different amount. Individually, each problem will cause some degree of invalidity—a difference from the mean. Collectively, the problems in our process will create greater variance—lower reliability—as some problems create bearings that are too large, and others that are too small.

According to quality management theory, any variation that appears unlikely to be due to random factors, by definition, has a cause. It becomes the engineer's responsibility to figure out what is wrong, to assign a cause to the error. There are two elements on our chart that would be defined as having an *assignable cause*. Both are marked with dotted ovals. At the lower right, we see a single data point below our −3 sigma lower control limit. The likelihood of being outside our +/−3 sigma control limits due to random chance is low enough so that we want to assign a cause to that event. The same would be true for a point above the upper control limit (not shown on this diagram).

Near the center of the diagram, we see seven sequential data points all above the mean. On a chart with +/− 3 sigma control limits, seven sequential points either above or below the mean are also evidence of some cause of bias. How did statisticians come up with that figure? By definition, there is a 50% chance that each point will fall either above or below the line. For any two points, there are four possibilities (both above; above-then-below, below-then-above, both below), each equally likely, with a 25% chance. The chance of two sequential points being either both above or both below is, therefore 50%, or 0.5. As we add more points, remaining above or below becomes less likely by a factor of two for each additional point. So, the formula for the probability P that n sequential items will all be either above or below the line is:

$$Pr(n) = \frac{1}{2^{(n-1)}} \qquad (17\text{-}1)$$

If $n = 7$, then $P = 1/64 = 0.015625$. This number is very close to the probability of one point lying outside $+/- 3$ sigma (0.027). As a result, it is about equally unlikely that seven sequential events will be either above or below the mean without a cause as it is that one event will be outside our control limits without a cause. (The actual calculations for the likelihood of a run of seven sequential points depend upon the sample size and are much more complicated. However, the principle is based on the equation above.) In both cases, seven sequential events on one side of the mean, or one event outside our limits, our statistics indicate that there must be some non-random, assignable cause to these events. It is up to our engineers to find that cause and develop a permanent preventative solution so that our manufacturing process will remain within our control limits.

Note that an actual process control chart would have hundreds—or perhaps millions—of data points, and a computer would be used to examine the data and find important information about assignable causes. In our example, production requirements have set the specification limit just outside our 3-sigma level, called our *control limit*, so our goal is to maintain process control within 3-sigma variation for our sample. Since we can't sample every item, we seek to keep all sampled items within a narrower range than that required for our population.

It is important to remember that we see data points from our sample set, and not from our entire population. One way to picture this is to imagine many other invisible points between the points we see. If we find one item outside our control limits, then there are probably many others, and some of them are defective, that is, outside our specification limits. If we find seven sequential items all on one side of the line, there are probably a few items produced in that time period that are outside our control or specification limits, as well.

We use statistical process control to define what changes are needed to our manufacturing or work processes. The result of these processes is our product. Because we do not sample every component, and because a single product will have specification limits on many variables and be the result of many processes, we may need very tight tolerances on many variables to create an acceptable product. It is a mistake to think that six sigma process control would give us only 3 or 4 rejects per million products. Errors combine in complicated ways to create unacceptable products. Extensive planning and analysis are used to define what are the most important variables to monitor and control.

At the right-hand end of our control chart, we see a plot of the variance, grouped by sigma ranges, turned sideways. (Look for the light gray squares.) If everything is going well in our manufacturing process, that will approximate a normal curve. Looking at the curve, we see a bias above the mean, and an extreme point below −3 sigma. This plot gives us a picture of our assignable causes.

A 6 sigma process control chart would be nearly identical to the 3 sigma chart, except that our control limits would be set to 6 sigma, and our requirements for the number of sequential items on one side of the mean having an assignable cause would be stricter. One interesting aspect of 6 sigma theory is the proposition that any non-normal variance must have a cause, no matter how small it is. If we can't find a cause, then we need to refine our measurements so that we can. Iterating process improvement with data from statistical process control applied through Pareto optimization, root cause analysis, and permanent preventative action, it is theoretically possible to reduce error levels until those levels approach zero.

SOMETHING EXTRA

How Good is Good Enough?
Just because we can improve quality and eliminate errors doesn't mean it is good for business. Once good quality management practices are in place, the more we improve our process, the more it costs to find and eliminate the next error. At a certain point, it may be cheaper to have a certain number of errors than it is to try to prevent them.

The solution is a complex process called analyzing the cost of quality. Perhaps it should be called analyzing the cost of error. If we can't fix a problem in manufacturing, but we can find all defects in QA before we go to the customer, then the cost of each error is either the cost of rework to fix the unit, or the cost of scrapping the unit, whichever we decide to do.

But what if we can't catch all the errors, and some customers receive defective products? Then cost of quality depends on the consequences for us and for the customer. There have been some famous cases—the gas tank location in the Ford Pinto, and some more recent issues where SUV tires suffered tread separation— leading to fatalities. In these cases, companies considered only their own legal liability, and not the lives of their customers, and they got into very big trouble. When lives are not at risk, then, below a certain number of defective products, it is cheaper to pay off on warranties and guarantees than to improve our process.

There is one other solution. We can turn lemons into lemonade. Our defective products may be useful products in some other forms. There are many examples of this. In ball bearing manufacture, a single manufacturing process makes precision ball bearings for high-speed motors and sensitive medical and engineering equipment, lower-grade ball bearings for roller skates, and bee-bees for b-b guns. Products that are too low grade for one purpose are sold as a cheaper item. I've seen lopsided ping-pong balls sold as cat toys, and defective carabiners (mountain-climber's hooks) sold as keychains.

Most interestingly, the LCD sheets that become flat screens for TVs and computer monitors have millions of transistors per square inch. In manufacturing, a few defective transistors are inevitable. Huge sheets are produced and tested. Whenever a large area (say 36″ by 27″) is perfect, it becomes a very expensive large-screen TV. Smaller defect-free areas are used for smaller high-resolution devices. Areas that have too many defects for high-resolution can be used for lower-resolution devices such as large-print displays. Smaller pieces make LCD watches. At the very lowest end, we have $1.99 LCD watches with just three numbers on them, which require a few dozen working transistors.

By finding uses for many grades of our manufacturing product, we make money from a product that would otherwise be considered defective. Then our goal is to generate the optimal product mix over time to satisfy the demand for each product. We may actually lower the manufacturing quality of our product periodically in order to increase production of lower-quality items, if demand for those items exceeds supply.

Quiz

1. Which of the following is *not* part of quality management?
    (a)   Engineering processes
    (b)   Statistical tools
    (c)   Quality analysis
    (d)   Defined process changes

2. Quality is ...
    (a)   Freedom from error
    (b)   Conformance to specifications
    (c)   Return on investment
    (d)   Reduction of error incidents

3. The goal of quality management is
    (a)   To reduce errors to the lowest possible level
    (b)   To bring errors to an acceptable level and maintain them at that level
    (c)   To identify and eliminate errors
    (d)   To maintain processes within control limits

4. Pareto diagrams are used primarily in association with:
    (a) Root cause analysis
    (b) Statistical process control
    (c) Zero-defect initiatives
    (d) Quality management

5. In a Pareto diagram, types of incidents are:
    (a) An ordinal list, arranged from highest count to lowest
    (b) An ordinal list, arranged from lowest count to highest
    (c) A nominal list, arranged from highest count to lowest
    (d) A nominal list, arranged from lowest count to highest

6. For a statistical process control diagram, which of the following is *not* *necessarily* an event with an assignable cause?
    (a) Seven sequential events above the mean
    (b) Seven sequential events below the mean
    (c) One event outside the control limits
    (d) A skewed variance plot

7. Which of these is *not* true? Six sigma is
    (a) A goal of fewer than 3.4 defects per million events
    (b) A measure of variance
    (c) An industry buzzword for a particular school of quality management
    (d) A replacement for Total Quality Management

8. Which of these is *not* a way of reducing the cost of quality?
    (a) Ensuring that all measured variables conform to the tightest possible requirements
    (b) Finding ways of selling, rather than scrapping, items that do not meet quality assurance requirements
    (c) Permanently eliminating the most common sources of error
    (d) Emphasizing prevention over reviewing and testing

9. Pareto's law is also known as
    (a) The six sigma rule
    (b) The 80:20 rule
    (c) The 20:80 rule
    (d) The law of the mean

10. You have reached the optimal level of quality when
    (a) Your help desk staff is as lonely as the Maytag repairman
    (b) You have reduced defects to the 6 sigma level
    (c) You have minimized the cost of quality, that is, the cost of error
    (d) All data points in your sample are within control limits

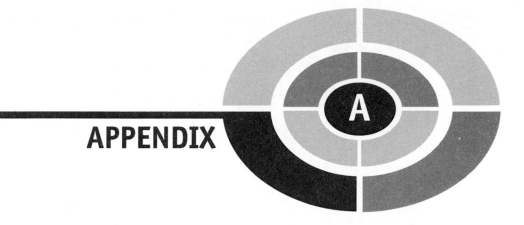

Basic Math for Statistics

One reason why some people have trouble learning statistics is that they find the math confusing. Our first challenge is that, when statistics uses numbers, it uses different types of numbers. The number line in Fig. A-1 illustrates some different types of numbers.

The simplest numbers are the counting numbers: 1, 2, 3.... These are called positive integers. If we add zero, we have the non-negative integers. If we add negative numbers, we have the whole set of integers, zero, positive, and negative. We use these numbers for counting, addition, and subtraction. The number line extends forever in each direction, because we can always add one to any number, making a higher number.

If we multiply and divide integers, we find that the answers to certain division problems fall between the integers. For example $5/2 = 2.5$, or $2\frac{1}{2}$. All of the numbers we calculate through division are called rational numbers. That includes all integers, and an infinite number of rational numbers between each integer. In fact, there are an infinite number of rational

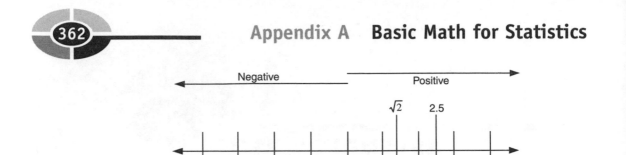

Fig. A-1. The number line.

numbers between any two rational numbers, as well. They are called rational because they can be created by division, and they are a kind of ratio.

There are more numbers on our number line, called irrational numbers. Some of them are found through questions like this: $2^2 = 4$, so $\sqrt{4} = 2$. What is the square root of 2 ($\sqrt{2}$)? It turns out that there is no rational number that, multiplied by itself, equals 2. So, $\sqrt{2}$ is an irrational number, approximately equal to 1.414. All of the numbers on the number line, rational and irrational together, are called the set of real numbers.

Things get interesting when we compare the scales used in statistics (introduced in Chapter 2 "What Is Statistics?") with the types of numbers on the number line.

- *A nominal scale* has no actual relationship to numbers at all. But we sometimes assign numbers instead of names, so that we can use statistics. We can sort a nominal scale in any order we want, because it is names (nominal) only, and not intrinsically ordered.
- *An ordinal scale* has some of the qualities of positive integers—counting numbers. There is no need for a zero or negative numbers. And each item is discrete, as the integers are. So we often use counting numbers for an ordinal scale. But there is one important difference. We know that integers are evenly spaced. We don't know that about the values in an ordinal statistical scale.
- *An interval scale* maps very well to integers, because it has both order and evenly spaced intervals. However, it has no definitive, meaningful zero point. As a result, calculating ratios on an interval scale makes no sense. (We know that two is one more interval than one, but we don't know that two is twice one.)
- *A ratio scale* has a meaningful zero point. Our data may be in integers for example, because we collected data by counting items—or it may be in real numbers. Our statistics will be in real numbers, both rational and irrational. The mean is always a rational number, as it is calculated by division of rational numbers. Other statistics, such as the standard

deviation, which is calculated with a square root, can be irrational numbers. Sometimes, statistics are rounded to real numbers or integers for convenience.

Learning statistics will be easier if you can work with basic algebra easily. If you want to bone up on your basic math, be sure to cover these topics: mathematical symbols; signed numbers; addition and subtraction; multiplication and division; fractions and how to reduce them; percentages; significant figures; properties of zero and one; factorials and exponents; binomials and binomial expansion; square roots; the commutative, associative, and distributive properties of mathematical operations; and methods for simplifying equations.

Where can you learn more about these topics? Start with the back of your business statistics textbook. If you want more, be sure to take a look at *Algebra Demystified* from McGraw-Hill.

APPENDIX

Answers to Quizzes and Exams

Chapter 1   1. c   2. b   3. a   4. b   5. c   6. a   7. d   8. c   9. a   10. b

Chapter 2   1. b   2. d   3. a   4. b   5. d   6. a   7. b   8. c   9. a   10. c

Chapter 3   1. b   2. a   3. c   4. c   5. a   6. d   7. b   8. c   9. d   10. b

Exam: Part One   1. b   2. a   3. b   4. c   5. a   6. d   7. c   8. a   9. b
10. b   11. d   12. a   13. b   14. d   15. a   16. b   17. c   18. a   19. c
20. a   21. c   22. a   23. d   24. b   25. c

Chapter 4   1. b   2. c   3. d   4. a   5. b   6. d   7. c   8. d   9. b   10. d

Chapter 5   1. c   2. d   3. a   4. c   5. d   6. b   7. c   8. d   9. a   10. d

Chapter 6   1. c   2. b   3. a   4. b   5. d   6. a   7. c   8. b   9. a   10. c

Chapter 7   1. c   2. a   3. d   4. b   5. c   6. d   7. a   8. d   9. c   10. b

Chapter 8   1. b   2. c   3. a   4. d   5. c   6. a   7. b   8. c   9. d   10. a

Chapter 9   1. b   2. a   3. c   4. c   5. a   6. b   7. d   8. a   9. d*   10. c

Chapter 10   1. d   2. b   3. a   4. c   5. d   6. a   7. b   8. c   9. d   10. a

Exam: Part Two   1. d   2. c   3. a   4. d   5. b   6. a   7. d   8. c   9. b
10. d   11. c   12. b   13. d   14. a   15. b   16. c   17. a   18. b   19. c
20. a   21. d   22. c   23. b   24. a   25. c

Exam: Part Three   1. d   2. b   3. c   4. a   5. b   6. d   7. c   8. a
9. d   10. b   11. a   12. c   13. b   14. a   15. b   16. d   17. c   18. a
19. c   20. d   21. b   22. a   23. b   24. c   25. d

Chapter 15   1. c   2. a   3. d   4. b   5. c

Chapter 16   1. b   2. a   3. d   4. a   5. c   6. b   7. c   8. a   9. d   10. d

Chapter 17   1. c   2. b   3. b   4. a   5. c   6. d   7. d   8. a   9. b   10. c

*Answer D for question 9 in quiz for Chapter 9 should read "None of the above,"
rather than "All of the above." The confusion results from unclarity in the text and
the first author apologizes.

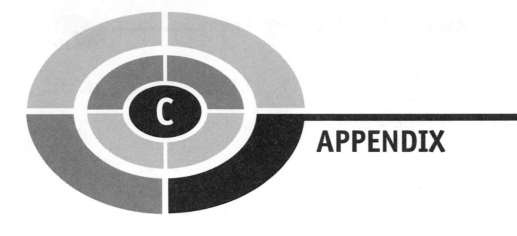

APPENDIX

Resources for Learning

We are constantly learning more about business statistics, and we want to share it with you. Rather than putting a lot of material here, and having it go out of date, we have created a web page for the book. Please go to *www.qualitytechnology.com* and click on *Books*, and select *Business Statistics Demystified*, or go directly to *www.qualitytechnology.com/books/bsd.htm*. On this web page, you will find a complete glossary for this book, introducing statistical terms for business, and a good deal more. We provide an up-to-date cross reference for the textbooks most commonly used for classes in business statistics, so that you can know which chapters of *Business Statistics Demystified* match particular chapters of the textbook you are using.

Also, we love to keep improving. Send comments or questions to *sid@qualitytechnology.com*.

Preparing for Exams

Our focus has been to help you understand statistics and use it to support business decisions. You probably want to pass your exams, as well. If you are looking for a book that provides drills, practice exams, and lists of equations to memorize, you will need to supplement *Business Statistics Demystified* with one of the workbooks specifically designed for that purpose. Several books are available. Two we would recommend are:

- *Schaum's Easy Outline of Business Statistics* by Leonard J. Kazmier, McGraw-Hill.
- *Flash Review: Introduction to Statistics* by Julie Sawyer and David Frey, Pearson Addison Wesley, 2002.

References

The following books were cited in or helped guide in the writing of *Business Statistics Demystified:*

Berenson, M. L. and Levine D. M. (1996). *Basic business statistics: Concepts and applications.* (6th Edition.) Englewood Cliffs, NJ: Prentice Hall.

Converse, J. M. and Presser, S. (1986). *Survey questions: Handcrafting the standardized questionnaire.* (Series: Quantitative Applications in the Social Sciences, No. 07-063.) Newbury Park, CA: Sage.

Cook, T. D. and Campbell, D. T. (1979). *Quasi-Experimentation: Design and analysis issues for field settings.* Boston: Houghton Mifflin.

Deming, W. E. (1986). *Some theory of sampling.* New York: Dover.

Huff, D. (1954). *How to lie with statistics.* (Pictures by Irving Geis.) New York: Norton.

Malkiel, B. G. (1973). *A random walk down Wall Street.* New York: Norton

Mosteller, F. and Rourke, R. E. K. (1973). *Sturdy statistics: Nonparametrics and order statistics.* Reading, MA: Addison-Wesley Publishing Co.

Rea, L. M. and Parker, R. A. (1997). *Designing and conducting survey research: A comprehensive guide.* (2nd Edition.) San Francisco: Jossey-Bass.

Thomsett, M. C. (1990). *The little black book of business statistics.* New York: AMACOM.

Weigers, K. E. (1996). *Creating a software engineering culture.* New York: Dorset House.

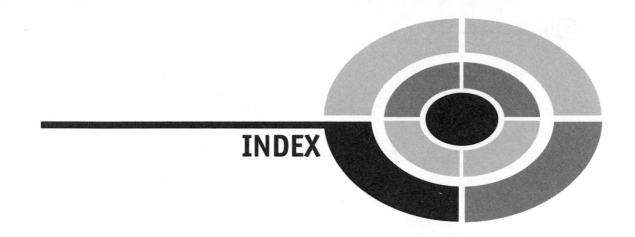

INDEX

ABOUT THE AUTHORS

Steve Kemp is a Research Assistant Professor at the University of North Carolina at Chapel Hill, where he received his Ph.D. in Psychology in 1993 from the L. L. Thurstone Psychometric Laboratory. He has taught Statistics at the Master's level at the Nursing school at UNC/CH. Steve does research in such exotic areas as neural network simulations of behavior, situational evaluations of behavioral simulations, and abductive reasoning. He is the inventor of the InSitu testbed (*www.InSituTestbed.org*). In previous lives, he has worked in data processing, software development, technical writing, advertising, and retail. This is his first book. When not writing or researching, he is a computer programmer and statistical consultant. Steve lives south of town with his cat, Eureka. Those interested in Steve's research or personal life can browse his Website: *http://www.unc.edu/~skemp/*

Sid Kemp, **PMP**, is a certified Project Management Professional and a leading expert in project management for information technology, with over 15 years of experience. He assists Fortune 500 companies and major governmental agencies in the deployment of new technology and in developing and deploying custom project management methodologies. Sid has trained thousands of first-time managers and team leaders in PM—in public, corporate, and government seminars. He has over 10 years of experience as a trainer and author of training programs and manuals. Sid is the author of two other McGraw-Hill books that may be of interest to readers of *Business Statistics Demystified*: *Budgeting for Managers* (2002) and *Project Management Demystified* (2004).